DISCOVER SWAGING

DISCOVER SWAGING

David R. Corbin

STACKPOLE BOOKS

DISCOVER SWAGING

Published by
STACKPOLE BOOKS
Cameron and Kelker Streets
P.O. Box 1831
Harrisburg, Pa. 17105

Published simultaneously in Don Mills, Ontario, Canada
by Thomas Nelson & Sons, Ltd.

To Katherine

Printed in the U.S.A.

Library of Congress Cataloging in Publication Data

Corbin, David R., 1944–
Discover swaging.
Includes index.
1. Handloading of ammunition. 2. Bullets.
3. Swaging. I. Title.
TS538.3.C67 1979 683′.406 78-22085
ISBN 0-8117-0497-1

CONTENTS

Introduction

It was obvious, years ago, when my brother Richard and I first got involved in supplying bullet swaging equipment to the world at large, that some day we would have to begin writing a book to show just how much could be done with swaging. There was no one at the time supplying information of any kind about the vast bulk of the field. All that could be seen were a few references to handgun bullets, usually very standard and rather dull designs hardly calculated to stir any real experimenter's interest—or references to ultra-precise rifle bullets for the benchrest game.

Rick and I were simply amazed that no one in the field of gun writing or publishing had ever tried to bring swaging information to the attention of handloaders for the great amount of versatility it offered between these two narrow fields. It was as if generations of sports fans had been told about both goal posts at either end of a football field, but had never seen or heard of the field itself. We soon learned that we were pioneering in a way, by trying to offer bullet swaging equipment of all kinds from one source, and that our business of supplying everything for swaging was really the only one of its kind.

Consequently, we were offering tools and supplies which few if any writers and publishers had heard of before, or for purposes which no one in the arms field thought would be of sufficient interest to warrant discussion.

All the information on swaging had previously been contained in chapters in benchrest or gunsmithing books and in a few articles and comments from time to time in the two narrow fields mentioned. As far as making bullets for hunting, for rifle calibers other than benchrest, for highly effective police loads or big game or metallic silhouette or—well, name any use of handgun or rifle that might require a special bullet designed to get more out of the gun than an ordinary, over-the-counter bullet—the techniques and equipment simply were not discussed. Rick and I knew that some of our products had been developed and died in the past, and that we were offering only improved versions of what many great gunsmiths and inventors had made years before. Our tools for making .224 and 6mm bullet jackets out of fired .22 cases were an attempt to make a standard tool available to everyone, not really an invention, since many great men of the past used similar ideas: Harvey Donaldson, Huntington of RCBS, Vernon Speer of Speer Bullets, and others.

We also knew that some of the novel techniques and ideas we had come up with would take years to reach the handloading public, since we had little money to promote and advertise to such a small market. Bullet swaging is done mostly by a small fraction of handloaders, who are themselves a fraction of all shooters. Shooters are a fraction of the total outdoor sporting goods buyers, and so it goes. There is a need greater than Richard and I can fill alone, as we typically run from three to six months behind in special orders for equipment, and are never in need of something to do around the shop. But there is less than a burning desire in every shooter's heart to make jacketed bullets at home, especially since there are far more shooters than handloaders, and only a handloader would really be interested in making bullets anyway.

So, in reality, this book came about for three reasons and to fill three general needs. First, it was written for the handloaders who either want to know more about swaging, or who are already into swaging and need a good, comprehensive answer book to help with techniques and methods. Second, it was written for shooters all over the world, in every facet of the shooting sports, who are curious about what this thing called swaging might be, and want to find out whether or not it even interests them enough to try it, once they get into handloading. It may help them to understand a little about bullets and how they can be made to do certain jobs better, even if it does not inspire them to try swaging.

Last, the book was written for the gun writers and editors who have not had the time (and what editor or writer does, these days?) to indulge in swaging experiments to any great degree, yet who would really like to have someone who is actually involved in the "hands-on" work of swaging offer some ideas and comments, provide some general background and specific ideas of what can and cannot be done within reason, and generally supply a source of information that is not secondhand or too narrow in scope, but general enough to apply to any part of the shooting game from airguns to muzzle loaders. The experiments in this book were actually done, over and over again. The dies and tools were worked out ever so painfully, with many mistakes and thousands of dollars in time and material thrown away as better ways came to light. This book is the summary of not only my work, and my brother's, but also of experience with innovators like Ted Smith, whose S.A.S. Dies eventually became the Corbin business, with many changes and improvements in equipment.

I have to mention that wherever possible, other brands and diemakers are used for illustration. But most of the time it simply isn't possible. There are not that many other people in the world at this time offering the tools. And most diemakers are specialists in benchrest work, where the dies are very costly and made to do extremely precise work. This rather precludes using any other brands, since there are none, to illustrate jacket drawing equipment, special swaging presses designed for automatic ejection, rifle dies made to fit the standard loading press in .22 and 6mm calibers, and many other special tools which are Corbin products. This book isn't meant to be a catalog or advertisement, and if there were any way I could prove good intentions by using some other maker's equpment for illustration, I would gladly do so. Every other diemaker and supplier of swaging equipment that we have met has been friendly and helpful to us, and we certainly would not want to give the impression that there is any personal or business reason why other manufacturers' products are not included.

The problem of being one of the few, or the only, supplier and maker of certain kinds of equipment for swaging is that the techniques and ideas that come to light in years of full-time work with swaging are hard to pass on to the public. Before I became well known for the present business, I did considerable free-lance writing for gun magazines as well as communication and business journals. I wrote about Ted Smith and his dies, since he and I had become friends. As long as I wasn't involved in business with Ted or with my own swage dies and tools, the articles were accepted and published. I could get information across to the public that had not been offered before. But as soon as I started selling swaging equipment, it was impossible to get an article in print about marvelous new techniques handloaders could use to make very superior bullets, using anyone's dies, even if my own products were not mentioned.

Never let it be said that the gun magazines or their editors are not looking out for their readers' interests! I understand and appreciate that it would appear to some readers, perhaps a majority, that the magazine was accepting some kind of public relations material from me if they did publish even very useful information regarding bullet swaging. I only wish the great majority of our discoveries and tests had been done before I became unpublishable! There is so very much that needs to be brought to the handloader's attention regarding bullet design, and making bullets at home. Not just because it is fun and the equipment for new ideas is being developed every day, but because it offers whole new dimensions of performance for many kinds of guns, changing the picture often presented in articles which deal only with factory or cast bullets. I cringe every time I read about low-velocity .44 Special loads in a light frame revolver not expanding, or the pocket autos being compared to revolvers and coming far down from where they might with the right bullets. (Often I have a dozen actually in my pocket, which turn the results of a current article completely around and make the conclusions reached invalid, but I cannot reach the people to say, "Hey! That isn't so if you just make a bullet like this, and here's how.")

Well, handloaders and friends in the shooting and writing fields, now at last I can say it. It took years, working nights and weekends, between making dies and running the business, testing the ideas and coming up with new ones. This book has been written and rewritten so many times that it would be an encyclopedia, if all the copies were bound and stacked side by side. Often I have had to change whole chapters, or to write new ones, as we found new and better ways to do old jobs, or discovered ways to accomplish things formerly considered too expensive or difficult. I could have also made this book twice as thick by writing it in the first person and relating all the interesting (maybe, to experimenters and machinists) work that went into developing each kind of tool or die, and telling anecdotes about the work we did, the people we met, and the often very funny experiences we had with other shooters and diemakers.

There isn't much fat and puffery included, though. I did include a little from B&A literature, which is of very great historical interest and hopefully is not out of place here, even though it isn't necessary just to understand how to swage bullets. And Nosler Bullets of Bend, Oregon, did a fine study of bullet nose deformation which involved the development of their Protected Point bullets, which really isn't swaging information so much as a very novel bit of information that could help swagers to determine a source of inaccuracy in their bullets and to correct it, or to work out their own version of the Nosler idea. The chapter on Hoxie bullets is included because they did exist, and I tried some of the ideas and found that among the humorous and extravagant claims were a few tidbits of useful and novel technique for making bullets a little better. Besides those diversions, there is textbook information that a handloader can turn to and use immediately: explanations of how the dies work, how to work out bullet designs that are practical, how to use the tools and techniques to develop bullets better than you can buy, and in general a guide to swaging that goes beyond making standard, run-of-the-mill bullets by rote and recipe.

The chapter on making dies is not intended to discourage competition—there's plenty of work for everyone, heaven knows!—but to put a little light of reality on the enthusiasm of those experimenters and hobbyists who read the book and decide immediately that they want to make some dies, whether to save money or to just have fun. Swaging bullets is easy and a lot of fun; making dies is hell unless you enjoy torturing yourself over vast amounts of slow, hard work with no promise of any success until the very end of the job. Then, the end result is about half skill and half chance; the slightest error destroys the die, since only a few ten-thousandths of an inch stand between success and failure. Perhaps, if the demand is great enough, Richard and I will write a book on diemaking that spells out the secrets and techniques of years of hard work—that in itself being the biggest secret—and at least gives a good machinist a fighting chance without all the false starts. This book is about swaging, not diemaking.

I hear Richard grumbling pointedly in the shop, next to my little office, right now about some work that needs doing, and the telephone is ringing again, interrupting me for the hundredth time (it seems) today. I have something like forty of those tricky Mity Mite presses to get out before too much longer, along with hundreds of dies that are at this moment in the form of bars of Timken Graph-Air, twelve feet long, looking greasy and sad and not at all like the highly polished gems they are going to be, if I ever get this writing done. I will go ahead now and put the heat treatment ovens on, set up the lathes, and get back to work on those dies. It has been a long, hard pull—this book—and I am glad it is done at last. But it needed doing, and I would go through it again, if necessary, to make sure that somewhere in the world there was at least one source of complete information available about swaging. I hope you enjoy it.

Chapter 1

Swaging: Gateway to Adventure

Bullet swaging is forming bullets under pressure, applied to the lead and jacket (if a jacket is used) by a punch and die set. Heat is not used. The bullet metal is made to flow like a viscous hydraulic fluid under the intense pressure. This is the fundamental difference between cast and swaged bullets.

A good deal of what has been written about swaging is bunk. The misconceptions include everything from mispronunciation of the name (it is not "swagging," like the swag a traveler carries over his shoulder on a stick, but is pronounced like "stage" or "wage") to stories told over and over again about performance in handguns, cost of equipment, time required to make good bullets, and the applications for handloaders. It is time to set the record straight. Few authorities in the firearms field have attempted to do so, and most of those attempts have not been written from wide enough experience to be truly useful to all potential bullet swagers.

Until now there have been no comprehensive books about swaging. There have only been articles about narrow aspects of the field. Each article seems to be presented as a comprehensive view of the field, when in fact the surface has hardly been scratched. There are more kinds of bullets, more styles and uses and techniques for making bullets at home, that are yet undiscovered both in the firearms press and by the general handloading public, than all the uses which have been written about so far. Two narrow fields have drawn all the attention in swaging, and even that attention has been slight. First, benchrest rifle bullets, because of the high precision and great control swaging allows, have been long associated with the high cost, slowness and tedium of swaging. Second, half and three-quarter jacketed pistol bullets of rather standard design have been written about, since this represents a popular and low-cost way to make jacketed bullets at home. About the rest—nothing.

Yet, the equipment is available for those who know where to get it. Custom tools can even draw jackets from common materials, reuse fired .22 cases or shotshell primers, turn one jacket caliber into another one with different characteristics, or allow the handloader to create built-in controls in expansion, penetration, and accuracy. The kinds of swage equipment available now are more varied than at any previous time in history, and the control they give to the handloader over the performance of the bullet is greater than any other factor in handloading! The very ratings of the various guns a handloader uses for different purposes can be radically altered by the use of unusual homemade bullets, bringing small pocket autos and low-velocity revolvers up several steps in defense or small game effectiveness. Bullets can be made which cannot be purchased anywhere, for any price, and will so vastly outperform any available factory ammunition or bullets available for reloading, that it is completely astounding none of these have found their way into articles, or made themselves forcefully known among handloaders.

There are a couple of reasons why swaging and its powerful techniques for bullet design have not been discovered yet, except for a handful of shooters. First, bullet swaging gives the handloader so much control that it almost eliminates the need to depend on factory supplies, and bullets are a very profitable item! Swaging dies and components have never been too profitable in comparison. Why should any bigger business try to promote a field that can hurt its own more profitable sales?

Second, bullet swage dies are unlike other reloading tools in that they are very difficult to make, time-consuming, and require a highly skilled diemaker full time. Bullet swage dies are far too critical in diameter, surface finish, taper control, roundness, and other parameters to be automated. There never has been, and never will be, a bank of automatic machinery turning out high-grade bullet swaging dies like toothbrushes or reloading dies. Each die is made by hand, lapped with diamond-using tools also made by hand to fit precisely with each other. The only people foolish enough to try to make a living at this are dedicated diemakers, and at this time there are perhaps six throughout the world producing any substantial amounts. One of them produces more than all the rest combined.

Some swage diemakers have been notorious for secrecy and have even closed their shops so that visitors could not see how the dies were made. Others have been more open about their techniques, because there really isn't any secret that twenty years of experience won't reveal. And, usually, that in itself is the biggest secret: years of hard work and experience, to get a feel for the operations and the hand work. No one can steal that. It has to be earned. That is why we are more than happy to present a chapter on diemaking in this book. It is too hard for the average machinist to want to attempt anyway. Using the dies is just the opposite; nothing, including casting, could be easier or safer.

With all the production in the world being limited to a handful of sometimes secretive diemakers, and most of these producers working

This .44 Magnum bullet was designed for fast, reliable expansion, for police work in city areas.

part-time just on special calibers of interest to them, such as benchrest rifle calibers, is it any wonder there are no floods of press releases, no books except those written by Corbin in recent years and the one old manuscript by Ted Smith published by Corbin, no information to speak of anywhere to be found? The real misfortune in this is that there are thousands of shooters who could benefit greatly from the kinds of bullets that swaging allows them to make at home. Both the low cost of some styles and the use of free jacket materials would let more shooters get in more practice. The superior performance of some designs would give off-duty policemen and home owners a bigger edge against criminals without danger to innocent people being any greater, as with the fast-expanding low-penetration designs. Multiple jacket rifle bullets would give every rifle owner the advantage of a two-part bullet that stays together under the most severe high-speed impacts yet still expands reliably even at extreme ranges. And so on, in nearly every field of shooting from benchrest rifle to muzzle loader, from defense handguns to air rifles, there are better swaged bullets one can make cheaply at home.

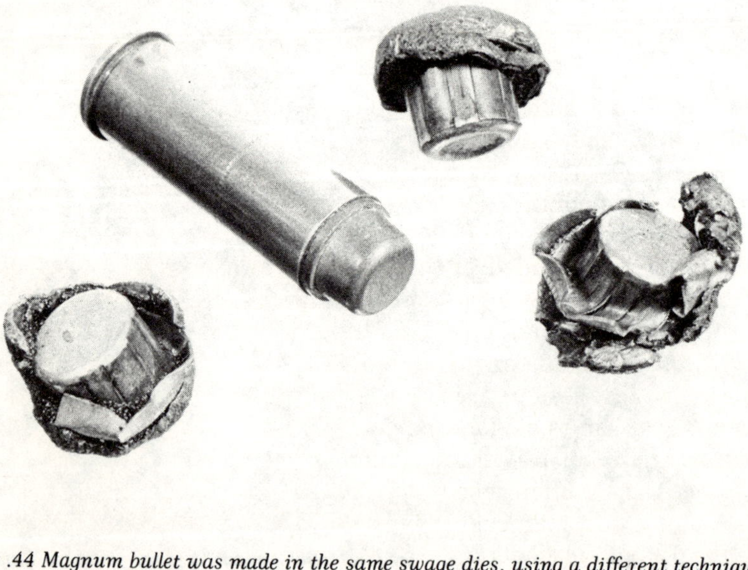

This .44 Magnum bullet was made in the same swage dies, using a different technique. It is for use on medium game.

But the real thrill of swaging is the process itself. It is fun and quick, not too hard to learn, and once mastered, it gives the handloader so much freedom to design bullets of his own construction. Some people will have trouble with swaging and will damage tools no matter how much is written, because it is a field requiring intelligent and careful work rather than brute force. The dies and tools are very forgiving of mistakes, when one learns how great the forces are within them. But to the person who does not like to follow instructions and learn to do things right, swaging can present a real problem with dies cracking apart and punches smashing into steel pancakes, as the pressures can become so great as to be almost beyond belief. Yet the effort it requires to create these powerful forces is very small. Most people are very excited once they discover how much swaging can do for their shooting. They are the dedicated handloaders, the experimenters, and nothing is ever quite good enough if there is a promise of something better. It is for them this book is written. They are the hard inner core of the shooting game, the legion of handloaders who have advanced beyond tinkering with the powder and the bullet weights offered by the factories, and want to actually make the one part that does the work, delivers the energy, and makes the shooting worthwhile. Swaging is for people who know handloading, and want to carry it to its utmost limits.

But one does not have to be an engineer to be a successful bullet swager. Some of the best designs were made by trial and error. One of the best bullet makers is a contractor, another is a salesman, and another is a welder. Some of these people are making such precise, unusually good bullets in odd designs that perform better than anything else they could get, that one would think they had spent years at the art. In fact, most of the good swagers we know have had less than six months' experience; some turned out outstanding designs that became commercially successful after the first month. They are all open-minded, and trust in their own ability to test, discard, and try again. They are not afraid to experiment.

Besides the thrill of producing unique bullet designs, the savings in cost for high-grade or high-performance bullets, and the fact that some of the best designs aren't even available from the factories, there is another reason to explore the swaging field. Some calibers are not available any other way. For example, the .20 caliber bullet is made only by hand. It offers good performance in rifles and at least one person has made a .20 caliber pistol based on a small bolt-action rifle like the XP-100. Jackets are drawn from .224 standard commercial jackets or else .22 short cases. Other calibers such as old European sporting rifle calibers like the 9.3mm are no longer made. They are easy to swage. Caliber has no barrier to the bullet swager, since the die diameter can be anything the diemaker will lap out.

There are two general classes of dies, the standard dies which may be offered "off-the-shelf" as stocked items, and the custom dies made to order. Standard dies offer little in the way of options for rifle shooters, but they do offer a very low price and high quality as a result of this standardization. Custom dies offer just about anything a person wants and can pay for. Prices go from slightly more than standard dies to several times as much, since a custom die may require the making of a great deal of tooling, which is used once and thrown away.

In handgun calibers, standard sets of dies are so versatile that they usually represent the best bargain and should be considered strongly. The difference between rifle and handgun dies is made more clear in chapter 4, but basically the handgun die set makes bullets which take their primary form in the operation of expanding the jacket and filling it with lead, while the rifle dies make bullets with the primary form controlled in the ogive-shaping operation, after the lead has been pressed into the jacket. A handful of different punches makes a vast array of handgun bullets from a single set of dies.

In all die sets, the weight of the bullet can be adjusted simply by adding or taking away a little of the lead that makes up the bullet's core. So, while each die set makes one fixed diameter of bullet, it can make bullets which are suitable for different kinds of guns of the same bore size. The versatility of swage dies is better understood after looking through chapter 4. But without question, a person can make many more weights and styles for a given investment by using swaging equipment, than he can by using casting equipment. And the velocity and performance range of those swaged bullets far exceeds what cast bullets have to offer. Swaging opens a gateway to adventure, and to a lifetime of expanded pleasure from handloading and shooting.

Chapter 2

Swaging Versus Casting

There's really no conflict between swaging and casting techniques. Both have their advantages and disadvantages. The choice depends on the purpose of the bullet. Casting is usually cheaper in price per bullet if for no other reason than cast bullets don't have the extra cost of a jacket involved. Of course, swaged bullets don't have to use jackets either. They do when the jacketed bullet is the correct design for the purpose. Equal cost for material results when both processes use only lead.

In cost of equipment, casting starts out by being cheaper for one particular bullet weight and shape. Casting involves not just the cost of a mold but also a lead pot, ladle, lubricator and sizer machine, die and punch set for the machine, and of course the extra time involved in melting lead, inspecting cast bullets, cleaning up, sizing and lubricating, and then the belling of the case mouths and the more careful seating procedure necessary to avoid damage to the lead bases. Even so, a person can start for about the same cost or a little less expensively in cast bullet making. If the bullet so made works well in the gun, and the handloader is pleased with the speed, accuracy, and performance, then casting is the way to go.

The advantages of swaging come into play when that same handloader who was pleased with his one weight and shape of cast bullet goes to purchase other molds and sizer dies, and starts experimenting with weights and shapes, speeds and terminal performance. It becomes obvious that every mold is fixed by the manufacturer for a specific weight and shape of bullet. There is nothing to be done except purchase another mold for each change. If a handloader desires an in-between weight or a different shape, or a basic change in bullet design requiring a jacket, double cup and hollow point design, or any other special changes no mold manufacturer feels is popular enough to include in the line, he can either place a special order at high cost and wait, or turn to swaging.

The cost of the swaging equipment and the length of time it usually takes to get good equipment made is about the same as a good custom mold or two. But the advantage is that the one investment in swaging dies and equipment will permit an almost unlimited number of different weights and styles of bullets to be made at will, with no additional equipment or waiting for special parts. That is, a set of dies to make a 150-grain .308 bullet will make the same bullet with a fair amount of variation in weight and nose styles, but with an infinite number of steps between the ranges of weight and degree of ogive. The handloader can fine-tune the bullet to perform the best possible way within the range, with no arbitrary steps or unavailable weights between any two given bullets within that range.

With casting, the handloader is stuck with precisely the one weight and style of bullet for his investment in a mold, and if another ten grains of a slightly longer ogive or a wider open point would be better after all, the choice is to spend the same thing over again for a different mold. Part of the trouble is that the handloader will never know if the bullet really is the best possible for the purpose because he cannot change the weight or shape in small increments, nor can he do a thing with the basic design. It is a lead slug, with or without a gas check, and the advantages of using various jackets and combinations of cups, hollow points, multiple jackets or cores, or other techniques of swaging are simply lost.

It is reasonable to say that there are at least ten different distinct variations possible with most swage dies, at the very least. That is, from heaviest to lightest bullet that will still shoot, and from the widest tip to the smallest tip, a person should be able to make about ten bullets that anyone knowledgeable about bullets would pick out instantly as being different. If a mold only cost fifteen dollars (and try getting any custom-made mold for this kind of money today!), it would take at least $150 worth of molds alone, not counting changes in lubricator or sizer dies or punches, to equal what can be done with a single set of swage dies. Of course, this is all beside the point if the cast bullet is just the right thing for the job. Shooters who mostly use cast bullets in target matches for pistols want only consistent performance at low cost and good accuracy at medium speeds, and can easily satisfy their requirements with a single good mold.

The performance of the bullets is really the key to whether a person should be considering swaging or casting as alternatives. Most people who shoot use both methods side by side. When higher speeds

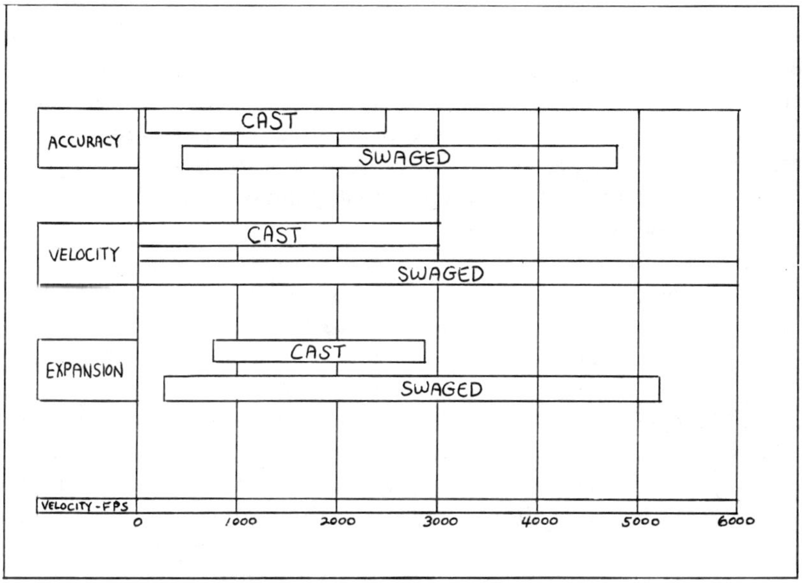

Chart comparing cast bullet and swaged bullet performance ranges.

and a greater control over bullet performance are required, swaging usually is the choice. A swaged bullet is good for the highest rifle speeds possible, if the jacket and design of the bullet are correct. A cast bullet usually starts limiting its own performance from fouling, deformation, stripping across the rifling, or failure to perform as expected on game, in the range that starts a little over 2,000 fps. This is very general, of course, but few will deny that cast bullets aren't often used in benchrest work and are almost never seen at speeds over 3,000 fps.

Performance of the bullet when it strikes is important to the game hunter, varmint shooter, and of course anyone interested in guns for defense. Lead is an ideal bullet material since it is dense and carries velocity well against the friction of the air. The more a projectile weighs for a given diameter, the more energy it can be given to use in overcoming air friction. This means it will slow down less as it passes through the air, and thus will cover the same distance in shorter time than a less dense bullet. Gravity pulls on the bullet all the time, regardless of how fast it goes, and the drop is the same for the same length of time, which is why the bullet that gets to the target first will strike higher, have less drop, and thus be easier to aim without so much guesswork.

It is easy to see how this works in a practical demonstration. A piece of drill rod can be thrown much farther than the same size wooden dowel, since air friction will quickly balance the smaller amount of energy a person's arm can impart to the lighter dowel. The steel rod will go farther, because more energy results from propelling this greater mass at the same velocity. Seemingly, the same effort is required to throw either, since the speed a person can move his arm is the limiting factor for velocity. But actually, more energy is expended in throwing the heavier projectile.

The gun works the same way. Although the more dense bullet can be driven the same speed as the bullet made of lighter material and will have more energy to overcome air friction, the limit is the pressure the gun can sustain (actually, it is the cartridge and not the gun that limits the pressure). If two bullets are driven the same speed, the more dense bullet will carry more energy, received from a higher pressure if the same burning rate results in the powder, or received from a longer burning time if the powder is matched to the projectile. The more dense the bullet, the shorter it can be for its weight. The shorter the bullet, the less spin is required to make it stable in flight. Any energy not used in spinning the bullet is energy available to propel it forward.

The sum of the matter is that more dense bullets are superior projectiles to less dense bullets. Now let's look at cast bullets and swaged bullets in this light. A cast bullet has no protection from barrel friction, powder gas, or air friction other than the melting point of the alloy and thus its hardness. In general, alloys used to increase the melting point also increase the hardness of the bullet and lower its density. As a bullet is driven faster, it can be made harder and still expand properly. Deep penetration on large game would require a hard bullet even if the velocity did not. Soft lead by itself, without the support and control of a jacket, would expand and thus give up energy too close to the surface of the animal tissue for a clean kill, in most cases. Hardness, to a point, is a virtue for this application.

Lightness, or less density, is not a virtue. If a pure lead bullet and a hard alloy bullet of the same shape and dimensions were fired at the same initial speed, over the same range, to the same game, not only would the lead bullet get there first, with more energy, but it would penetrate more deeply into the target provided some way could be found to keep it from expanding right away. It is not just air friction, but any resistance, that a more dense bullet carrying more energy will overcome better than lighter bullets, or bullets of the same weight which are made of less dense materials.

Putting a jacket on the pure lead bullet solves the problem nicely. The jacket keeps the bullet from fouling the barrel at just about any speed desired, keeps the air from heating the lead and deforming it, and keeps the bullet from expanding until it penetrates the desired amount. At the same time, the bullet can be more dense, and the jacket can be tailored so that expansion is controlled to just about any degree desired. This solves another problem with cast bullets made as hard as possible for high-speed work. Such bullets may become brittle, from too much hardness, and shatter when they strike a bone rather than expanding. Proper use of the jacket over the soft lead core can prevent this.

At the other end of the velocity spectrum, hardness is not a virtue. In order to keep from leading a revolver barrel, the bullet usually needs to be hard enough so that it does not expand as easily as pure lead. At lower speeds, such as those found in the .44 Special or the pocket automatics, cast bullets offer rather disappointing performance since the choice seems to be between bore fouling from lead, or a bullet that just won't expand and stop quickly. Penetration is seldom desirable with these kinds of bullets, as they are used against small game or for defense. Yet, penetration is the main characteristic of a hard cast bullet.

The jacketed swaged bullet can simply be made so the tip is left quite wide and with a good deal of exposed soft lead. Or a hollow cavity can be put into the nose. The jacket will keep the bullet from fouling the bore; yet the soft lead will expand easily on contact with any substantial target material. Performance of small handgun calibers can be greatly enhanced by the use of the proper kind of swaged bullet. Here, a cup base is normally used in conjunction with a cup or cavity nose, something difficult to accomplish with cast bullets. The cup base helps the lower pressure seal the bore by expanding the base of the bullet very slightly on firing. Thus more of the available energy is used to propel the bullet.

A cast bullet does have lower barrel friction than a swaged bullet,

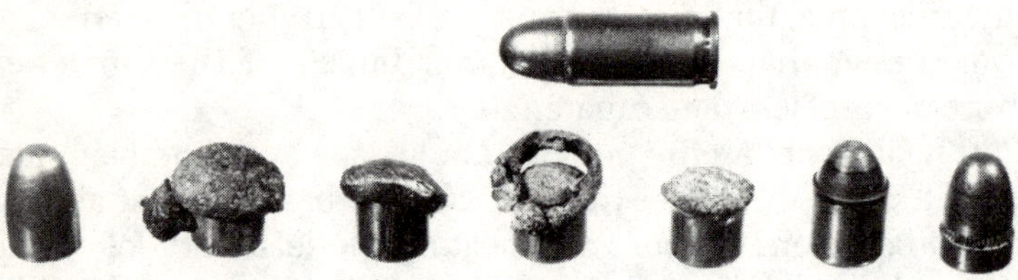

The performance of small handgun calibers can be greatly enhanced by the use of swaged bullets. Here a .32 ACP cartridge is shown with factory bullet next to a swaged bullet. The same swaged bullet recovered from four different target mediums, next to the factory bullet after firing into the same targets, shows superior performance of the homemade bullet.

and a slight increase in velocity results with the same powder charge. This is more than balanced in most cases by the improved performance of the swaged bullet. Those who feel swaged bullets are only good for high-speed work have not experimented with the proper designs for low-velocity jacketed bullets, since the results speak for themselves. It is true that some target work done with jacketed bullets at low velocity proves a few guns and loads do not give best accuracy with a jacketed bullet at target speeds. The shooter who is primarily striving for accuracy should try both kinds of bullets and then decide. It is quite possible that a cast bullet will do a better job. But it isn't a fixed rule that every gun has to shoot poorly with jacketed bullets, certainly not with every design of a jacketed bullet. The swage die gives a handloader a chance to find out through careful experiments just how to make a jacketed bullet perform well at low speed. Factory jacketed bullets are usually not designed for this, and much of the problem has been with the use of factory bullet designs.

In summary, it is likely that a swaged bullet can offer improved performance for defense and for big game hunting, both at high speeds and at low velocity, due to the ability of the handloader to tailor the expansion characteristics of the bullet to suit the purpose. There is one other large advantage to a handloader: time. Time is usually short enough for hobbies. Anything that makes it possible to turn out shootable ammunition faster makes it possible to enjoy more actual shooting time. Casting by virtue of the need to melt lead and to handle the bullets several times before they are loaded takes considerably more time than swaging.

For example, a pot of lead probably takes a half hour to melt. This is lost time. The mold needs to be brought to temperature. The first few bullets probably will need to be remelted. As casting progresses, the pot must be fluxed and the mold kept at a fairly even temperature. When the casting is done, the bad bullets are tossed back into the pot following inspection, and clean-up time begins. Nothing can be put away until it is cool. The bullets are still not ready to shoot. They must be sized and lubricated, which means handling each bullet individually one more time. The sizer and lubricator must be filled with lubricant, the correct punches and dies put into it, and the assembly adjusted. Excess lubricant has to be removed from the bases of the bullets before they can be loaded.

While the handloader who casts his bullets is waiting for the pot to melt, the handloader who swages bullets can probably get out his swage dies, cut the cores, and swage the bullets. The advantage of swaging is that there is virtually no clean-up process compared to casting, no time for the lead to melt, assuming that cut cores or a stock of precast cores is made up, and only one step in handling the bullet prior to loading it. With semiwadcutter-style bullets there is only one pull of the press handle to turn a core and jacket into a finished bullet, ready to be loaded. In short, swaging is faster than casting. It is significantly faster, so that spur-of-the-moment reloading of only five or ten rounds is practical, and the bullets can be made any way within the range of weights and styles of the die set. Changes can be made from one bullet to the next, just by selecting core lengths and adjusting the die up or down in the press.

For the handloader who protests that he wishes to cast both cores and bullets, and feels that cast cores would be just as slow as cast bullets to make, there are two offsetting factors to consider. First, the core molds for swaging are not necessarily the same kind as those for bullets. Being a cylinder without grooves or ogive, the core mold cavity can be fitted with a plunger and mounted to the bench, ejecting the cores up from the cavity rather than having to be opened each time. Even a four-cavity mold can be made in this way, with individually adjustable plungers run by one bar. Since the mold mounts to the bench and is equipped with a long-handled sprue cutter, one hand operation is possible. There is no mallet to pick up, and pressing on the bar or plunger ejects the core from the mold without tapping any pivots. Such a mold is obviously much faster to operate than a bullet mold.

The other advantage is that cores can be cast up ahead of time and still not be frozen, like a cast bullet, into one style or shape. The core can be swaged into many different forms, to suit the needs of the moment. A stock of cast cores, of three or four different weights, will make dozens of different kinds of bullets, while a stock of cast bullets is exactly that: a fixed supply of a certain kind of bullet. The advantage of keeping a stock of cores of different weights available for instant use is that the time can be short for reloading, yet at any given time a supply of the exact kind of bullet wanted for the day is only a few moments away. Certainly the time for casting has to be considered, and it is only a little less than the time for making cast bullets. But the choice of when that time can be taken, and how much versatility is left after it is taken, is much more broad with swaging cast cores than with stocking cast bullets.

Of course, even a strict comparison of all the time involved in casting cores and then swaging them into bullets with casting the bullets and preparing them to be loaded would prove the swaging process is faster. This is because the ejecting type mold is quicker than a bullet mold, saving time in the actual casting operation, and the cores do not have to be perfectly filled or without any surface wrinkles in order to make good bullets. This results in fewer rejects and more production per hour. The core does not have to be run through a sizer and lubricator, but it does have to be put into the swage die with the jacket and swaged. The time is about equal for either process. Lubricant has to be wiped off the bullet in either case, except that for swaged bullets the lubricant is a water-soluble, light fluid instead of a heavy grease. In the actual steps of reloading, the swaged bullet is of course a little faster. So the total time for a swaged bullet, ready to shoot, is less than that for a cast bullet, ready to shoot.

This is only a valid comparison when bullets are made by swaging that can be made by casting. There are no bullets suitable for some purposes that can be cast, such as high-velocity rifle bullets. These have to swaged. Also, some kinds of performance levels such as extreme expansion at low speeds or controlled expansion at high speeds without leading the bore are simply outside the range of cast bullet capabilities. It is not fair to casting to compare swaging time on these kinds of bullets. Within the limitations of the cast bullet, the swaged bullet that substantially duplicates the performance is a little faster to make, even with cast cores. With cut wire cores, swaged bullets are much faster. Lead wire is more expensive than cast lead cores, but many people prefer to trade a little higher cost per bullet for the extreme speed of manufacturing the bullets.

In the realm of muzzle-loading bullets, swaging has a great advantage over casting. Besides being much faster and just as cheap, certain kinds of bullets can be swaged that are all but impossible to cast. These include the hollow-point Minie-style bullet with a deep cup base. Casting a bullet with a cavity in both ends is a fairly difficult task. Swaging one is as simple as making any other kind. It requires only one die and one stroke of the press. On the other hand, swaging a round ball is difficult and casting one is easy.

For paper-patched bullets in either muzzle loaders or cartridge guns, a swaged bullet offers a fine and fast solution. The smooth-sided, unlubricated shank of the paper-patch bullet is ideal for the swaging process. Two-part bullets for slug guns and for muzzle-loaded breech-loading guns (as with the Pope system) can be made in swage dies. A hard lead and a soft lead alloy are used to form the two sections of a bullet, and are swaged together to make one bullet. The admonitions against using hard lead alloys for cores is still true. But the dies and punches for hard-alloy bullets are designed differently and are not to be used with jackets. Pressure is kept within reason, so that even the harder alloys will work. Only soft lead should ever be used in a jacketed bullet.

In fact, one of the oldest uses for swaged bullets was the slug gun, an early form of benchrest rifle which achieved accuracy unrivaled until modern times. In the realm of high precision, there is hardly any question that swaged bullets will outperform any cast bullet, at least at rifle velocities. Suffice it to say that benchrest matches are won with either factory benchrest quality bullets, which are swaged at the factory, or hand-swaged bullets made by the shooter or a custom bullet maker. Cast bullets aren't noted for their presence at these matches, where everything depends on precision. Giving the cast bullet everything due it, the open-minded shooter would still have to take a hard look at swaging and concede that in just about every direction it extends performance beyond the possibilities with melted lead.

Chapter 3

Swaging Equipment

A general outline of swaging equipment a person might desire, and the minimum needed to make any sort of given bullet, is fairly simple. There isn't a great deal of accessory tooling required. The basic tools of swaging are the press and dies.

A press for swaging is ideally suited if it has about a two-inch stroke and a horizontal ram so that the handloader can see both the punch and the die easily. Special presses have been made from time to time that feature horizontal short stroke rams and are used only for swaging. Usually, special dies and punches that fit only these presses are also provided, and other advantages such as automatic ejection and floating alignment of the punches and dies can be incorporated.

For the smaller calibers and those which don't require a great deal of pressure to form, or rather, have such small total volume or area that the correct pressure can be created without much strain on the press, a normal reloading press can be used. The punches slip into the ram of the press like a shellholder, and the dies screw into the top as any reloading die might. Generally, it is best to use a heavy-duty type of press for swaging. Some of the standard light linkage C-presses will work reasonably well, but a multiple leverage system is better. The stronger O-type press is desirable.

Turrets and special presses may not work with certain kinds of swaging equipment. There is about as much pressure in the smaller calibers against the press parts as when full-length resizing a rifle case. Turrets tend to work loose, and in all slotted ram presses used for reloading, the thin top part of the ram can tend to be pushed down. Although the press frame can stand just about any sort of suitable force the leverage can provide, the ram top is not designed to sustain pressure against it. Larger rifle calibers and heavy handgun calibers require either a modification of the reloading press, by shortening the stroke, adding a fixed ejector pin to the frame, and replacing the ram with a solid top design, or a special press designed from the start just for swaging use.

A handloader who is going into swaging is best advised to purchase either one of the top-of-the-line reloading presses, if the caliber he desires to swage is suitable for the standard slotted ram type of punches and dies, or to go ahead and get a special swaging press for making bullets. It is very convenient to have a reloading press that can be left for reloading, and to have the little bullet factory complete in itself, ready to make bullets on a moment's notice. Generally, the specialists who do convert reloading presses for swaging with their dies will recommend a press or even supply it with the die set. For ultraprecise bullets, the press is usually reworked for better than standard alignment of the ram and top.

The dies used for swaging are described in detail in chapter 4, but basically there are four dies, and one can make certain bullets in just a single die. The most simple die is a straight-walled die with a hole all the way through it and punches fitted to both ends. This die makes the basic form for a core swage, a core seater, and a lead tip former. Another kind of die has a semiblind hole lapped into it, which is the shape of the bullet. This is a point-forming die, and is also called an ogive-walled die since the die walls carry the ogive shape. It has a tiny hole at the apex and a very tough ejection pin that presses the bullet out. A set of dies can be from one to four dies, each with at least two punches for the top and bottom.

The number and kind of dies desired depend on the bullet form. A bullet that is a straight cylinder with a little step and then a lead nose, like most factory handgun bullets in the so-called ¾ jacket style, can be made in a single die. A core swage die is used to make the lead core weight and form more consistent for accurate rifle bullets, and an ogive-walled or point-forming die is used to bring the jacket into a

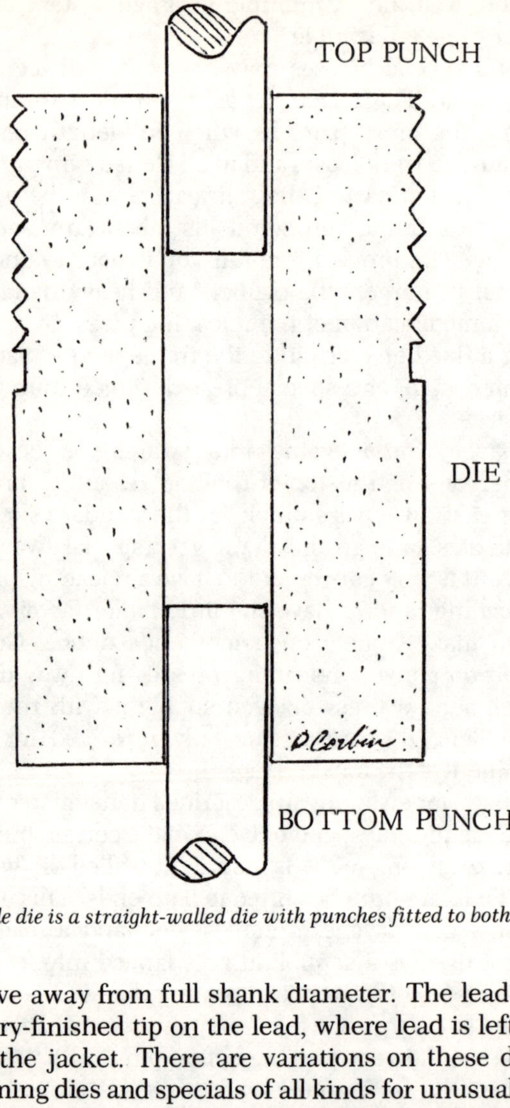

The most simple die is a straight-walled die with punches fitted to both ends.

smooth curve away from full shank diameter. The lead tip die simply puts a factory-finished tip on the lead, where lead is left extending by the end of the jacket. There are variations on these dies, including boattail-forming dies and specials of all kinds for unusual shapes.

The handloader doesn't have to know what kind or number of dies he needs in order to obtain a set to make a given bullet. Just a description of the bullet and either a request for jackets or a sample number of them will be sufficient for the diemaker to determine what is required. After the dies arrive, the handloader will want to know which die performs what operation. Quite often a handloader may specify dies that won't do the job specified at the same time, or may

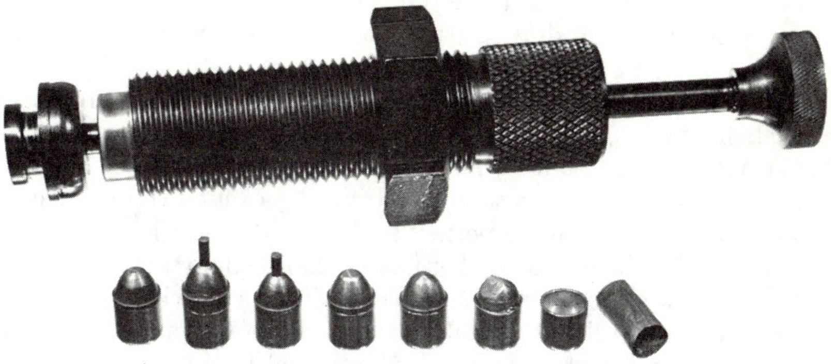

A practical straight-walled die made to fit standard reloading presses, along with the bullets it produces.

order dies that are not really necessary; thus it is best to order by either reading the available material from the diemaker first, or by simply ordering according to the kind of bullet and letting the diemaker put together a quote or a die package that will handle it.

One thing that is rather important is the jacket. Usually it is best to either send samples of the jacket which will be used with the dies and punches, or else to order jackets at the same time to use with the dies, from the diemaker. Bullet jackets are not all the same. Some are thicker than others, many have internal tapers, and the material or hardness may vary. Dies that work well with one jacket might not make the same size or shape of bullet easily with another. The problems are usually solvable but can be eliminated entirely when jackets and dies are properly matched at the beginning.

The accessory tools for bullet swaging might include jacket redrawing dies for certain calibers. A draw die is used to reduce the diameter of a jacket or, sometimes, a bullet. It is an open die through which the component is pushed. A narrow throat in the die does all the size control and a polished taper section does the gentle reduction.

Bullet-swaging dies, on the other hand, never reduce the diameter of a bullet but must always make it a little larger in each step. Bullets are not swaged down. This is a contradiction in terms. Swaging is cold-flowing the metal to fill the die cavity, expanding it by pressure to conform to the wall shape of the die and punch ends. Drawing is pushing a metal component through a ring die to reduce it in diameter. Just as one does not draw a part "up," one does not swage a bullet "down."

Draw dies are used to make jackets from fired rimfire cases, shotshell primers, and copper tubing. They are used to take all the taper out of swaged bullets for use in machine guns, autoloaders, or short-necked cases. A draw die can bring a jacket down one or more calibers or make it a more perfect fit for an odd-caliber bullet. Long, thick-walled jackets in a smaller caliber can be drawn from standard jackets of a larger caliber. A draw die also uses a matched punch that fits the kind of jacket material and thickness for which the set was designed. Sample jackets are needed for special draw dies.

Core molds provide a convenient way to make the lead cores, as long as caution is used not to cast regular alloy lead that might be used with cast bullets. Swaging dies are made to operate with soft lead only. Hard alloys have no advantage in a swaged bullet with a jacket and have the disadvantages of lower sectional density and more fragmentation instead of proper mushrooming. Also, they are very hard on the dies and punches because they require more pressure to flow. As long as the lead used is soft, core molds provide a good alternative to lead wire.

Lead wire is a component used for the cores of bullets. It is available from many diemakers, and usually comes in either sticks or 25-pound spools. It is cut to correct length (which means weight) for the bullet desired, using an adjustable core cutter or lead wire cutter. A stop pin on the cutter is set to limit the weight of the cores, and the handle is pulled to chop up a length of wire into accurate cores for swaging. Lead wire is available in all calibers, and need not be extremely close in diameter to work, as long as it will slip into the core swage die or jacket. If it is very much too small, the length required for a proper weight bullet will be excessive, and it will fold before it can be compressed into the jacket. This is about the only serious restriction on minimum diameter.

To eject bullets from a die, a mallet is usually used. An ejection or knock-out punch in the top of the die is tapped to push the bullet out. To speed up this operation and reduce wear and shock on the punches, a mechanical ejector of some sort may be utilized. These vary from frames that attach to the ram, reaching over the top of the press and contacting the knock-out punch, to multileverage "power ejectors" with their own handle, like a miniature upside-down press. Such an accessory is very convenient for regular reloading press type dies. The special presses often incorporate ejection on the ram backstroke, making this accessory unnecessary.

A cannelure tool can be used to put a factory cannelure ring or groove around a lead swaged bullet, for lubrication, or on the jacket of a jacketed bullet. The cannelure serves as an expansion control area, with moderate success, depending on the jacket and velocity. It also offers a crimping groove for hunting bullets subject to some force in feeding or carrying. It is necessary on heavy recoiling revolvers and on many autoloading pistol bullets. Cannelures can't be swaged in the bullet because of the difficulty of removing the bullet from such a die. Special multiple-part split dies can do this job but are not practical for home use.

Lubricants are required in every swaging step, to act as a buffer between the die wall and the bullet. Special high-pressure lubes made just for swaging are available. Other mixtures can be made at home, following the recommendations of the diemaker. Anhydrous lanolin, either by itself or mixed with castor oil, is a favorite lubricant for swaging. Ordinary case lubes are seldom suitable. Most lubrication is applied simply by having a film on the fingers that handle the components.

A good micrometer is a handy accessory, but it can also spell trouble if taken too literally. A micrometer is at best a device for approximation to the nearest thousandth, where bullet work is normally done with electronic comparators and optical comparators and gauge blocks doing work accurate to a few millionths. Use a micrometer as a tool, not a standard. Micrometers can and do get out of adjustment. They can and do read differently for different people depending on operation and interpretation. The bullet swager should use a micrometer as a gauge to tell if cores are being seated about the same, to see if a jacket is larger or smaller than the finished bullet, and as a comparison tool, not as a final word on exact size of any of the steps.

Extra punches and special ones made to bell jacket mouths, to seat cores longer than the jacket, to form special bases or cup noses—these are all add-on items that a swager might want to have for special techniques of bullet making. None are very costly. A punch usually is made to fit either a jacket interior or the I.D. of a certain die. A lead slug swaged in that die is a good gauge to send to the diemaker for fitting a punch, second only to having the die itself in hand. A jacketed bullet is not the best size indicator, as the jacket will tend to spring back slightly to original, preswaged size when pressure is removed.

Most of the accessory items offered by diemakers are explained in the literature each firm publishes, and other than the ones described, few of them will be necessary in order to start making good bullets. Accessory tools can be added later, if and when the need arises. The basic choice is the system of swaging itself: the kind of dies and press.

Chapter 4

The Swaging Process

In order to be consistent in talking about swaging, certain definitions and concepts need to be agreed upon. The definitions in the glossary of this book are pointed out as a guide. A bullet is composed of either lead alone, or lead and a skin of copper alloy called the jacket. The lead, cut or cast to length, or after it has been seated in a jacket, is called the core. This is the solid filling for the bullet.

A projectile is an object in flight, following the laws of ballistics and having a trajectory. A bullet can be a projectile when it is fired from a gun, but it really isn't strictly correct to speak of a projectile coming out of the swage die. The jacket of the bullet may be called a cup, but the term cup is often used for pistol jackets because they are shorter and look more like a drinking implement. A cup is also one of the first stages in drawing a rifle jacket; so the term also applies to unfinished draws of jackets at one or more stages. Let us use the term jacket for finished jackets and reserve the word cup for describing partial draws and concave nose or base shapes that are less deep than wide.

Cores can be cut from rolls of lead wire or cast in molds. Jackets can be drawn from fired .22 rimfire cases in some calibers, from fired shotshell battery cups (part of the primer), from copper tubing, or, more usually, purchased in lots of 250, 500, or 1,000 minimum size packages. Most diemakers have jackets for sale, or can recommend a supplier. A jacket is always undersized for the caliber before it is swaged. It is supposed to be that way, so it will expand outward and then spring back slightly to grip the core.

There are two basic forms of dies. The first is a straight-walled die which has a hole right through it at the same diameter. The second kind has the shape of the bullet lapped into it. A straight-walled die can be used as a core-seating die, a handgun semiwadcutter-style bullet die, or, with the addition of small bleed holes, a core swage. It also can be used for a lead tip-forming die, although in this capacity it is merely a guide and the punch does the entire job.

The most simple die is a piece of steel with a hole through it, and a pair of punches fitted to either end. If a piece of lead were put between the punches and squeezed within the die cavity, as with a vise applied to the two punches, the lead would first buckle and then flow to fill the entire cavity. Once enough pressure had developed inside the die to cause the lead to flow into each part of the cavity so that no room was left inside, any excess pressure would simply act to expand the die and shorten the punches.

By stopping just as the lead touched the die walls, one would be able to make an unlimited number of perfect cylinders of lead. Getting them out of the die would be simple. One of the punches would be removed, and the other one tapped on through the die. The process would be slow in a vise; so a method of holding one punch fixed a short distance into the die cavity would be better. Then the other punch could be used to push the lead in, and the fixed punch would act as the top for the cavity. To eject, this fixed punch would be driven down.

This is accomplished by making a cylinder of steel for the die, and screwing it into an adapter of sorts, that has a solid head to it. The punch is also fitted with a head and a smaller hole is put in the head of the adapter. Enough space is left in the adapter so that the punch can be the length of the die cavity. The length of the head on the punch determines how far into the cavity the punch will project when it is back as far as the adapter permits.

The hole in the adapter top lets one put a rod against the punch head and tap it down to eject the lead. The bottom punch is separate from this assembly. By threading the adapter and screwing it into the top of a reloading press and fitting the bottom punch to the press ram, a simple method of applying even pressure is obtained. So far this

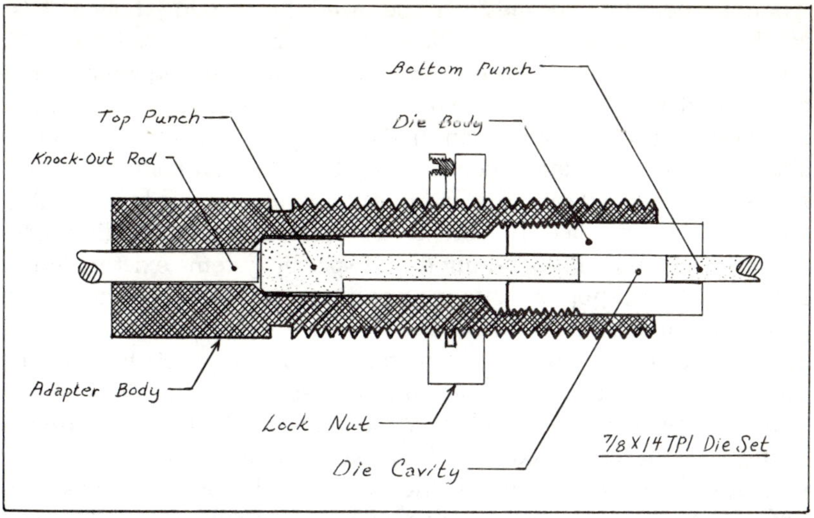

The parts of a die set made to fit a standard reloading press are illustrated in this picture of a core-seating die.

system can only make solid lead cylinders. The next step is to put a bullet nose on one of these cylinders.

The punches have been flat on the ends up to now. By reaming a cavity in one of the punch ends, a diemaker can produce a die and punch set that will create semiwadcutter-style bullets. The cavity in one of the punches, usually the bottom punch since it is separate and easy to change, is made in the shape of the bullet nose. Of course, the edge of the punch can't be made paper-thin unless very short punch life can be tolerated. So the bullet that this system makes will have a little step between the cylinder part of the bullet (shank) and the nose itself. The nose can be any shape, from round to conical to the Keith style. But there must be that step between the nose and the shank. The practical size of this step is between 0.040" and 0.010". It is not a bad feature, except that it does preclude the use of a jacket longer than the shank. Any bullet formed in a single die, with a step shoulder, will be considered a semiwadcutter style regardless of the actual nose form. This is because there is no real difference in equipment or technique in swaging any other nose shape on this kind of bullet. It is simply a matter of selecting a proper punch with the desired cavity.

To reverse the process, one can machine a dome or projection out of the punch. Then instead of a nose, the bullet would have a cup or depression. This is often used for cup nose or cup base bullets, and when carried to extremes can be used for making hollow base Minie-type or airgun pellets. The same die set is also used, then, for flat or cup bases, black powder-style hollow bases, or the whole range of semiwadcutter-style noses. There is nothing to change but a punch. And both ends of the bullet can just as easily be formed with different shapes at the same time, such as a cup-based hollow point. The more frequently changed punch should be the one more easily accessible, which is usually called the bottom punch in the regular loading press system. One should bear in mind that special systems exist which do not really have a top or bottom since they operate on the horizontal plane, but the same principles apply.

The jacket is swaged along with the lead core simply by dropping the lead into the jacket, and placing the assembly in the die. The lead must be longer than the jacket by the amount which will be used in forming the nose. The edge of the jacket can't be brought up against the edge of the nose punch, in other words, or the jacket would simply fold. A true boattail bullet can't really be made very well this way, either, since the punch would have such a thin edge that it would soon break or wear away. A rebated boattail, where the edge of the punch can make a step in the jacket, can be swaged in this simple die set.

The jacket must be smaller than the die cavity. If it requires much pressure to force the jacket into the die, then it will be very hard to eject it after applying tons of force to make it conform to the die walls. Jacket materials are somewhat elastic, and tend to spring back ever so slightly in the direction that would make them the same size they were before swaging. If they are larger than the die before entering, then after swaging they will want to stick to the die walls. If smaller, they will try to release the tight grip on the walls and can be ejected. Jackets as small as a whole caliber size under proper diameter will swage up to make fine bullets with the right technique.

The technique for expanding a very small jacket to fit the next larger caliber is to press a tapered punch into the mouth very gently and bell the mouth. The mouth of the jacket should then be a gentle press fit to the die bore. When the long lead core is pressed into the jacket, it will swell and expand the jacket to fit the die. If the jacket remained undersize at the mouth before the core was seated, then lead would simply bleed past the mouth and a poor seal would result, giving a bumpy, undersized bullet.

To make a bullet without this step shoulder, one in which the jacket can actually be brought around the ogive or curved part to any desired distance, a second die is used. The lead is seated in the jacket exactly as before, using either a hollow point (projection) or flat punch against the lead. A cylinder results that has a lead tip, with or without a cavity in the end. This is the seated core. It really is the seated core and jacket assembly, but for simplicity one refers to the seated core and assumes it is understood that it is seated in a jacket.

The seated core is ejected and put into the second die, which is the ogive-walled or point-forming die. A point-forming die is made with the exact shape of cavity as the finished bullet. Of course, it isn't necessary to press the seated core all the way to fill this cavity; so various degrees of ogive can be applied to the bullet by adjusting the press to push the seated core in more or less far. The seated core is put into the point-forming die with the lead toward the die, which is the opposite of the straight-walled die's operation. A flat or domed punch can be used to press the seated core into the point-forming die. Then comes the interesting part: ejection.

Ejecting a bullet from a point-forming die is done with a very small pin punch that is held within the adapter just as the top punch was held in the straight-walled die. A spring holds the punch out of the way of the forming bullet tip, and is compressed when the bullet is ejected. Some special systems use no springs, but have a provision for positive mechanical retraction of the pin. This requires a special press, or modifications to a reloading press. The bullet was pressed in to this die under very high pressures, and with a full-diameter punch. To remove it requires breaking away from the die walls and using a punch that can only be as large as the smallest point on the tip of the bullet. The smoothness and taper of the point-forming die must be very carefully designed to permit this to occur.

Naturally, every step has been preceded by handling the core or jacket with fingers having a film of suitable lubricant on them, or by this stage the bullet will be stuck fast in the point-forming die and require some emergency measures to remove it. Drilling or heating the bullet out of the die is a sure way to destroy the die, and one would save much time by simply throwing it away in the first place. Very simple and effective procedures are available for removing stuck bullets, the most effective of which is to return the die to the maker. Other techniques are discussed later. Prevention is the best one, but nearly everyone experiences a stuck bullet at one time or another.

The bullet sticks when the small ejection punch drives into the lead instead of pressing the bullet out. This occurs more often when the very tip of the bullet is extremely small, so that the pin must be very tiny. When a tiny pin is used, there is little area upon which to press and the pressure necessary for ejection of the bullet is applied to a small section of lead. Penetration is more common as the tip size, hence pin size, diminishes. A common pin size would be 0.061 inch in the .224 caliber, and 0.081 inch in the .308 caliber. These tips are sufficiently small and yet allow good ejection.

In making lead-tip bullets, the seated core is exactly the same as with any of the semiwadcutter handgun bullets or the very simple lead cylinder in the first example. The lead extending beyond the jacket is sealed off by a punch that is a gentle slip fit into the bore of the die. But a jacketed rifle or handgun bullet which has the jacket completely covering all the lead—in other words, a style called the "open tip" or "open tip full jacket"—used in most match bullet designs, and some special high-velocity handguns, requires that the core be seated shorter than the jacket. This means that a core must be put into the jacket, and a punch fitted not to the bore of the straight-walled die, but to the inside of the jacket.

Some jackets are straight and have the same thickness from top to bottom. In these, a punch of one diameter could be used for any weight of bullet. Most jackets, however, have a definite taper to the thickness. They grow considerably thicker at the base. This means the cavity within the jacket is shaped like a tapered plug. The core-seating punch must be made close to the right diameter for the size of jacket cavity

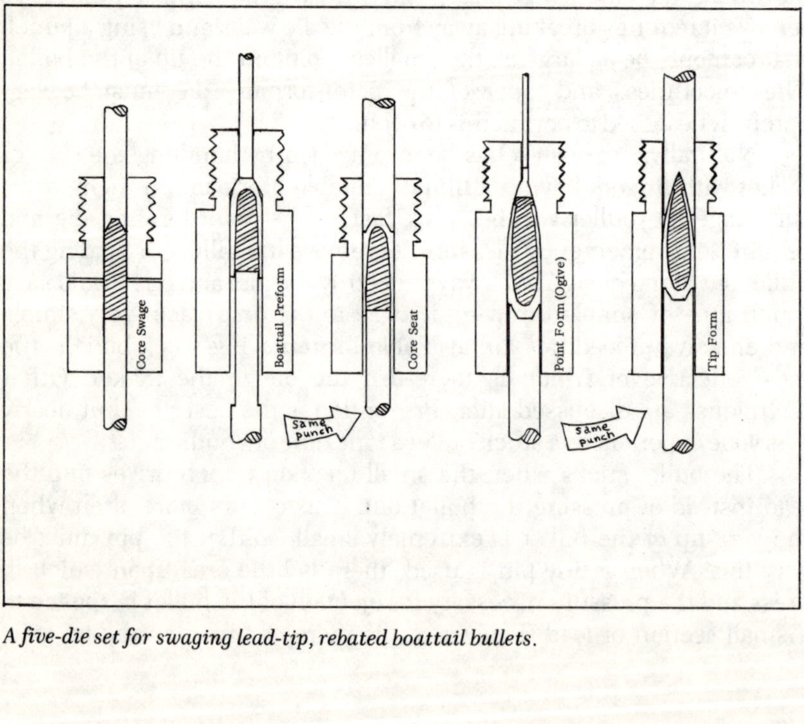

A five-die set for swaging lead-tip, rebated boattail bullets.

where the lead will finally come to rest. Different weights of lead will come to rest after seating at different points along the jacket wall. A heavier bullet will therefore require a punch that fits near the mouth, where the cavity is larger. This requires a larger punch than would be used for a lighter bullet in that jacket.

Also, different kinds of jackets have different thicknesses even at the mouth. A very heavy jacket used for big game might be 0.050″ or

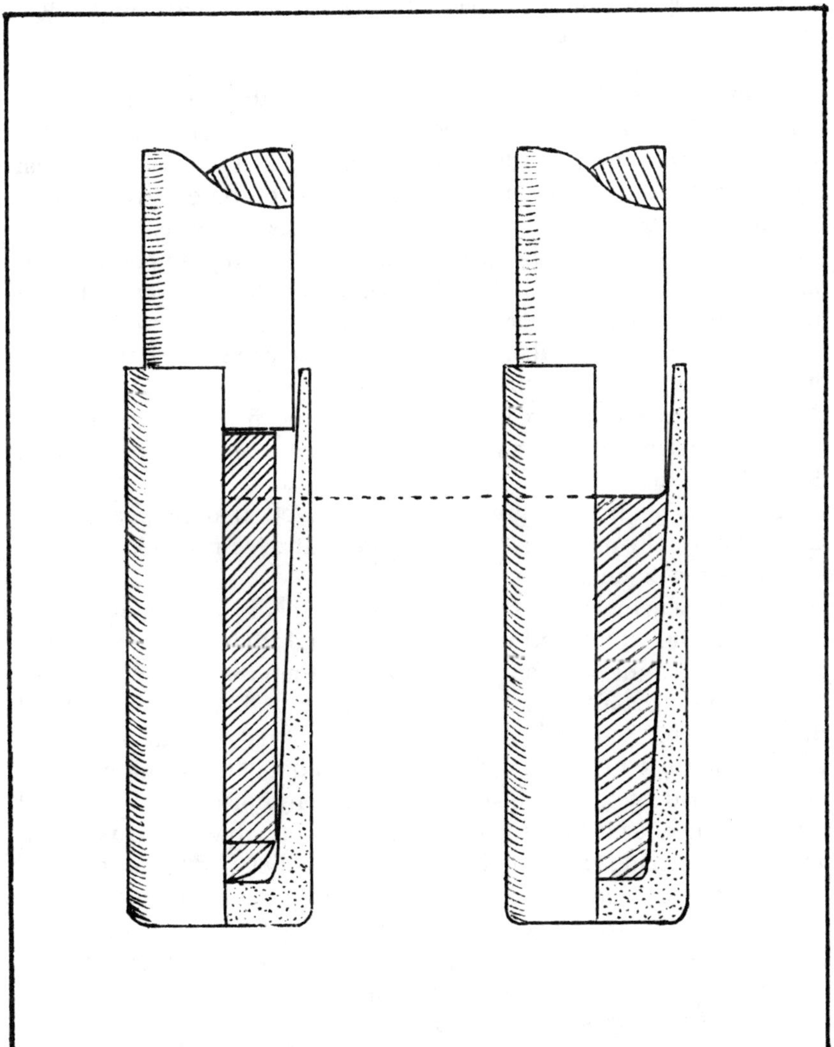

A properly fitted core-seating punch.

more thick whereas a varmint bullet might have a jacket of only 0.015″ thickness near the mouth. The punch must take this into consideration. The swager, that is, must select a punch that will fit gently without scraping into the jacket to be used, and at the point where the weight of the core will come to rest. Fortunately, there is some latitude in punch diameters and a little lead bleedby or scraping does not hurt, provided it is very small. The gentle taper of most jackets permits a wide range of weights to be made using only two or at the most three punches in a given caliber and jacket.

Whenever a die and punch set is made to be used with a point-forming die, the cavity size is made so that the seated core will be at least 0.0002 inch and not much over 0.001 inch smaller than the bore of the point-forming die. This straight-walled die is usually just called a core-seating die, since that's the real purpose. Any straight-walled die can be used for core seating, but some are used primarily for semiwadcutter bullets. The diameter difference is very important, and one should not order a .357 caliber semiwadcutter die set and then later decide to use it as a core seater for a point-forming die in the same caliber, unless the diemaker knows it. Then he can provide the proper difference in diameter in the point-forming die. If it turns out that the straight-walled die is already at precisely the desired size, then a slightly oversized point-forming die can be made and a matching ring die to bring the bullet back down to finished size supplied. This is effective only on the very small differences involved for proper ejection and is not usually recommended for a procedure to make, for example, next-size smaller bullets from factory bullets.

The bullets ejected from the point-forming die have been pressed by the ejection punch and will have a flat tip. If they are the type with the jacket brought past the core (open tip), then the pin will probably bear on the ends of the jacket and have no bad effect on the looks of the bullet. The lead-tip bullets with the small flat tip can be further finished in a lead tip-forming die. This is another of the straight-walled kind, but made with a bore slightly larger than finished bullet size. It acts as a guide to position the bullet against a punch with a cavity in it. This cavity is made to match the ogive of the bullet, and brings the lead to a point. Very gentle pressure is used. Too much would force the edges of the punch against the jacket and make that undesired step in the jacket, just as with any other straight-walled die and punch set.

If the bullets are for home consumption and not for resale, the tip can be finished when the bullets are seated, simply be reaming the bullet-seating punch with the proper shape of cavity. The pressure of seating in a firmly sized case neck will be enough to form up the lead tip. Actually, the very tip is a cosmetic improvement only, since the lead will probably be battered flat by recoil upon firing. This is only the amount of tip beyond the 0.081 to 0.061 inch diameter, and amounts to a very minor part of the ogive.

This is the basic swaging process, from a straight cylinder to a lead-tip bullet. Cavities, full-jacket military styles, multijacket designs, and many other techniques including the forming of boattail bullets can be done with simple technique or tooling changes, or the addition of only a single extra die set.

Chapter 5

Bullet Swaging Systems

A system of bullet swaging means a particular kind of die set with press and punches, punch-holding method, and technique of operation unique to that system. It is important to understand that there are several systems for swaging bullets by hand, and that each may have certain advantages. Usually, the tooling and dies, and the presses, do not interchange from one system to another. Some systems are offered only by one maker. Some are more or less universal designs but may be restricted to a special press or modifications to a regular press. A few work with the regular reloading press, unmodified.

One of the more basic systems is the so-called soft swaging, upswage principle of operation using two dies as a minimum to finish a jacketed bullet with the ogive formed partly or completely in the jacket. This is the system put forward and refined to near perfection by the early Biehler and Astles (B&A) bullet swage equipment. The chief competition in a different method of forming bullets came from R.C.B.S. (Rock Chuck Bullet Swage), and these two primary systems of swaging were the leading contenders for the precision swage die field at one time. The R.C.B.S. system used a single die that came apart to remove the bullet. The B&A system used two dies, and the bullet was formed first by making it a perfectly filled cylinder (seating the core in the jacket) and then by forming a nose on the cylinder in a second die. The R.C.B.S. system used much higher pressure to both seat the core and form the point at the same time, and was considered a "hard-swaging" method—"hard" referring to the pressure and not the ease of operation.

There were many other early makers of bullet swage dies, especially of the one-die take-apart type where the ogive was made entirely in one-half of a die and the shank in the other half. The Banta, the Hollywood, the Gibbs, and other similar two-part, hard-swaging systems all used various methods of trying to speed up the inherently slow operation of removing the bullet and reassembling the die. One of the major drawbacks of this kind of system is that it is slow to operate, and another is that alignment has to be extremely good between the two parts of the die, or a ring will result around the bullet. The ogive can be formed out of line with the bullet shank if there is wear or sloppy work in the fit of the two die parts.

The Pearson was one of the first of the single-piece, two-die systems of swaging offered for resale. The Easton was another early system using the soft swage method. The one-piece, two-die system has the advantage of perfect alignment between ogive and shank of the bullet, plus much greater speed of operation than the take-apart one-piece die. Primarily because of the high precision and attention to details of bullet and die accuracy, and the development of refinements in the various die parts and methods of securing the upmost precision in the bullets produced by their dies, the B&A swages proved themselves over the two-part dies and today's bullet swaging equipment consists almost entirely of the two-die, one-piece system of swaging.

Although these are two of the basic methods of swaging, there are many systems even in the two-die, one-piece category. The basic swage die for handgun bullets can use a single die, this being merely a core-seating die but with a nose cavity machined into one of the punches. The die can be fitted to an adapter that will screw into a regular reloading press. The top punch, which normally forms the bullet base, will be held in the adapter while the bottom punch, which forms the bullet nose, may be fitted to a button that slips into the press ram. The punch then fits just like a shellholder into a regular reloading press. The kind of ram slot, or even whether or not the particular reloading press uses slotted rams for shellholders, will determine which presses can be used with this system.

At this writing, the shellholders made by R.C.B.S., Pacific, and

C-H to fit their regular reloading presses are more or less the standards for all swage die sets that are made for use in "regular" reloading presses. Adapters or modifications to the press may be necessary to use what is commonly called a "regular" reloading press die set for swaging in any other style of press. A unique twist on this system is the L.L.F. system, where the die itself is fitted with a slot to slip into the ram of the press. The punch is held in an adapter in the top of the press, and the primer arm is used to hold what would normally be the top punch for the die.

A single die of the one-piece type may be capable of producing either a half-jacketed bullet, or both half- and three-quarter-jacketed bullets. There is usually a small bleed hole provided either in the nose-forming punch at the very center of the nose cavity, or else in the side of the die itself. This hole allows for variations in the amount of lead used to form the bullet. Once a certain weight of bullet is being made, the press ram is passed "over center" or as high as it will go while the die is adjusted so that this maximum position of the punch into the die will just form up the desired bullet weight. Variations in core weight would mean variations in the pressure and possible damage to the die and punches on heavier bullets.

The bleed holes provide an avenue of escape for the excess lead, and thus reduce the chances of swelling the die cavity if a few extra grains of core are included in some of the bullets. The location of the bleed hole will determine if a half-jacketed bullet, or both half- and three-quarter-jacketed bullets can be made in that system. A jacket covering all the shank of the bullet, as in a three-quarter-jacket style, would also cover the bleed hole in a die with the hole in the die cavity wall. This kind of die then would be restricted to the use of jackets which were shorter than the shank portion of the core. This kind of bullet is usually called a half-jacketed bullet.

If the bleed hole is in the end of the nose-forming punch, obviously it does not matter how long or short the jacket is within the die. This system is the more versatile of the two. It does not provide for automatic shearing off of the bleed lead, which forms a little wire or pipe on the end of the bullet nose. The side-bleeding die does shear off the lead when the bullet is ejected from the die. Clipping off the little lead wire is not much of a chore and can be done with the fingernail; so the punch-located bleed hole is a more popular choice.

If a standard reloading press is to be used with the die set, and the dies made to fit into the press top without modifications to the press, then the maximum size of die can be only about ⅝ inch in diameter. This is the size that will fit into an adapter to hold the top punch and still have enough metal in the walls to hold a ⅞ × 14TPI cut. Even if the entire die were formed of expensive die steel, machined as an adapter and die in one piece, the die walls become rather thin after anything larger than about .357 caliber is taken out of the center. Bullets up to .45 caliber have been made in dies of this size, but it is more likely that damage can occur from excess pressure.

Another problem exists with large calibers, or long rifle calibers which require more pressure than soft-nosed handgun bullets, in regular unmodified reloading press systems of swaging. The slotted ram simply is not made to withstand much force on the thin lip above the slot. While this ram could be made stronger by forming it from a hardenable tool steel instead of soft screw stock, the warpage and cost would be well above the limits necessary for reloading purposes. Thus, in rifle calibers, some diemakers restrict their production of dies for unmodified reloading presses to the .224 and 6mm calibers in rifle bullets, and to about .357 caliber in handgun bullets. This is not because the reloading press doesn't have sufficient leverage or strength of frame. It is because the ram itself is not suitable for pressure on its slotted top section, and can be damaged by forces necessary for larger calibers.

To solve the problem and make rifle calibers and handgun calibers in all sizes available for hand swaging, the slotted ram can be replaced with a solid top, threaded, or otherwise equipped ram to hold the die or the punch. The head of the press may be bored to a larger thread size, or if an adapter is already included in the press top for larger diameter dies, then the dies are made to fit. If the die fits the top of the press, bullet ejection is done by pressing down on the top punch either by a hand operated ejection punch or some sort of mechanical ejector unit. If the die fits the ram, then the press needs some slight modification in the form of a hole through the frame, below the die and in line with the vertical slot in the ram. A pin is put into this hole, which goes through the ram slot, and against which the internal punch rests. When the ram is lowered, the pin holds the punch from going back with it, and the bullet is pressed out of the die automatically.

Either system requires a new ram for the press, which must be made to fit the links or leverage pins and the bore of the press. The auto-ejecting system requires the press frame itself be worked upon, so that either the customer's press must be sent to the plant for modifications, or the customer must modify his press and put the parts into it himself. Of course, another alternative is that the press be supplied new with the dies, already modified and set up for swaging.

All reloading presses commonly found today are operated so that the ram moves up and down, in the vertical plane or at just a slight angle. The stroke of the press is almost universally around four inches. These features are necessary for handling reloading operations, as the powder would run out if the ram moved in any other plane, and the longer cases need the full length of stroke. Unfortunately, these two features are not particularly good for bullet swaging. All that is needed for even the longest small arms bullets is about a two-inch stroke. The rest is wasted motion, and worse, the extra stroke cuts almost in half the amount of power that can be generated as the ram approaches the top of the stroke. The stroke is gained only by sacrifice in mechanical advantage; one is traded for the other, and there is no logical way to demonstrate a system which can gain both without trading off one or the other. This is just basic physics.

The ideal bullet swaging system would probably not be with a vertical ram operation, because one cannot see up into the die if it is mounted mouth down, and feeding in cores, jackets, and bullets is more difficult. If the mouth is up, and the die placed in the ram to secure this, then the punch is more visible for protruding from the top of the press downward. But still, as the press does do its work near the very top of the stroke, the alignment of the punch to the die mouth is not as easily seen as if the ram moved horizontally.

A special bullet swaging press with a short stroke, and with provision for the ram to move in a horizontal plane instead of up and down, forms the basis for more than one top-quality swaging system. The early B&A press, of which very few were ever produced, used this method of press operation. The Mity Mite System of today uses it. And countless experimenters have surfaced over the years who have made their own special swaging presses in just this way. Such a press cannot be used for reloading, because the powder would run out of the case and only very short cases would be able to fit between the ram and the press head. However, if one is interested in the very finest way to design a press for swaging alone, strong consideration should be given to the horizontal, short-stroke ram.

Equally high quality has been achieved by some diemakers who purchase the best-quality standard reloading presses and then rework them so that the press does have a short stroke (the linkage pivots are moved closer together by making new holes or replacing the assembly entirely, and the alignment of the press head and ram made more nearly perfect by reboring and honing the casting). Sometimes special new rams of larger diameter are installed, and usually a solid-top ram of some sort is used. The dies can be made to fit a ⅞ × 14TPI hole, or perhaps the press head will be bored and threaded to a larger size. If the die is made to fit the ram, there is no particular limitation on its diameter, since it can be larger than the ram if it has but a small portion made to fit threads or other fastenings to the ram.

Usually the cost of the better systems, which have refinements like floating punch holders or close alignment presses, self ejection of bullets, and short stroke, is much greater than those which fit regular reloading presses. There is not much point in dwelling on refinements in accuracy on a system which is basically limited by the way the press head and ram align, by the method of holding the punches, and by the kind of shooting the customer who does not desire to go all the way with high-accuracy presses will be doing. To produce expensive benchrest precision dies for a regular unmodified reloading press would be an exercise in futility, for it would be unlikely that the system would produce bullets to the potential of the dies.

This is not to say that regular reloading press systems are inaccurate. It is a matter of degree. Competition benchrest shooters and those who feel that only the best is good enough would not be wise to consider regular reloading press swaging equipment. The rest of the shooting world could probably do quite well with such equipment. The key is probably use. If one is likely to demand benchrest precision from the bullets, or high enough speed of operation to make the venture potentially profitable, then the special presses and their dies are a wiser choice. If one would be quite satisfied with accuracy of the degree found in better factory bullets, which is quite high, and is not overly concerned with production speed while maintaining high accuracy standards, regular reloading press dies are a good decision. The factor of available calibers in each system also controls the decision.

Chapter 6

Presses for Swaging

From time to time, presses have appeared on the market that were said to be made for both swaging and reloading. Today we still see swaging mentioned in the advertising as a possible use, usually with such terms as "powerful enough even for swaging." Likewise, articles continually mention how strong a press is built by stating emphatically that it might even be stout enough to swage bullets.

All of this is only partly true. Any press that doesn't fall apart when resizing a .308 Norma Magnum case is plenty strong enough to do certain kinds of bullet swaging. And on the other hand, there isn't a reloading press on the market that will stand up under swaging use on large rifle calibers with a regular slotted shellholder type of ram.

Almost every reloading press sold today is strong enough, as far as the pins and the frame, to handle light-duty swaging such as the making of .22 and 6mm open-tip bullets in good dies, or making most of the smaller handgun caliber bullets. Just for the sake of a little extra margin, we ought to use only the C, O, or H type of presses. The Lyman turret press and the Bonanza Co-Ax press are too nice to even take a chance with, besides having other features that don't particularly make them ideal for swaging. It's hard to break the frame of nearly any of the semisteel presses available today. Cast iron frames, depending on the metal and the quality, can snap off under just about any strain if they are flawed.

The R.C.B.S. Rockchucker and even the little Junior are perfectly adequate for small caliber bullets in good dies. The Pacific presses, the CH presses, and all the others of good quality and reputation have strong enough frames. It is not the frame that limits the reloading press for swaging. It is the ram.

A standard reloading press with shellholder-type bottom punch and a ⅞ x 14 die for swaging pistol bullets.

The ram on nearly every reloading press is slotted for the primers to fall out, and slotted the other way for a shellholder. This means that there is a rather thin top on the ram, and with all those slots most of the manufacturers have decided that just plain screw stock is plenty good enough of a steel. Rams with that many cuts in them have very thin sections next to very thick ones, meaning that the loss would be very high in warpage and cracking if a tool steel were to be used, then hardened. A good tool steel ram would add a lot of cost, even if it were possible to make one out of nonwarping air-hardening steel with reasonable cutter life.

The slotted ram will work for .22 and 6mm calibers if the dies are reasonably good quality. We have used a Pacific Multipower for years for all sorts of swaging, testing, even force-fitting of parts. The only damage is that the edge of the ram bends down from time to time, and has to be pried up. Under normal swaging of lighter calibers, the presses on the market today are up to the task.

The presses with light leverage, rather than a hefty compound

The B&A horizontal ram swaging press (obsolete).—Photo courtesy of P. H. Sydney, Penfield, New York

lever system, are still plenty strong enough but they take a little more muscle power to operate and are not quite so convenient. We have used an R.C.B.S. Junior press in demonstrations of swaging just to show that it does work and without too much force, either, but even so, the more potent presses are easier to use. One press that is made for swaging, and yet has the slotted ram which has been hardened, is the big CH Champion press. But even so, a long and large rifle caliber such as a 180-grain .30 caliber bullet is quite a strain on the top of the ram.

For this reason, we do not believe that it is wise to use a regular loading press with a standard slotted ram for swaging calibers above

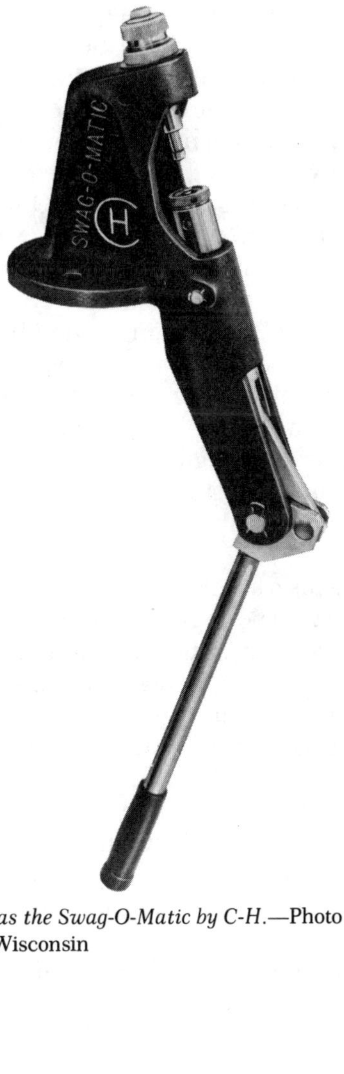

Another obsolete press was the Swag-O-Matic by C-H.—Photo courtesy of C-H Tool and Die Corporation, Owen, Wisconsin

6mm in rifle-length, jacketed bullets. By changing the ram and using a different system of holding the punch in the ram, or by holding the die in the ram, just about any of the better loading presses can swage bullets without any problems. The ram to use would be one with a solid top, threaded to take a floating punch or to hold the die. These would have to be made to order for the press.

Another answer was started by the famous B&A Dies firm, which was more of a hobby than a business. Biehler and Astles did much of the pioneering die work with accurate bullet swaging, and at one time only the R.C.B.S. and the B&A systems were considered worth serious attention from quality-seeking shooters. Both firms are long since out of the bullet swage picture, but while the R.C.B.S. A-2 press came about as an offshoot of this bullet swage activity, a more specialized press was developed by B&A just prior to their discontinuance.

The B&A press used a system of floating alignment, a horizontal ram, and was made just to swage bullets. The regular loading press, or the presses that are supposed to make both bullets and reloads, have a stroke that is nearly twice as long as desirable for making bullets. The B&A had a stroke nearly ideal for swaging. Only a few of these fine presses were ever built.

Another press that is built to operate with a horizontal ram, a short and extremely powerful stroke, and with a system of floating alignment much like the early B&A design, is the Mity Mite. This press is the one which we now produce for any caliber over 6mm and which is recommended for any of the commercial or benchrest work with swaging. We will not dwell on it here, other than to explain why any press is suitable for short handgun calibers and both .224 and 6mm open-tip rifle bullets, but almost no loading press with a standard ram ought to be considered for calibers much larger and longer than the 6mm.

The Mity Mite holds a special punch in a holder that permits both axial and angular float. The regular loading press with slotted ram permits a crude degree of axial float due to tolerance in the slot, but very little angular float. As the press springs under pressure (all presses spring just a touch, and come back again when pressure is relieved) the loading press puts a strain on the bottom punch. The punch does tend to align itself with the die bore if there is a long enough cavity before the bullet starts to form. But the Mity Mite system is much more positive and accurate.

Almost all loading presses have the operating handle offset so as to permit a full half-circle swing with access retained to the die. This offset places a stress on the press and a rotational force on the ram which is not conducive to the best accuracy and alignment of a punch

The Mity Mite press.

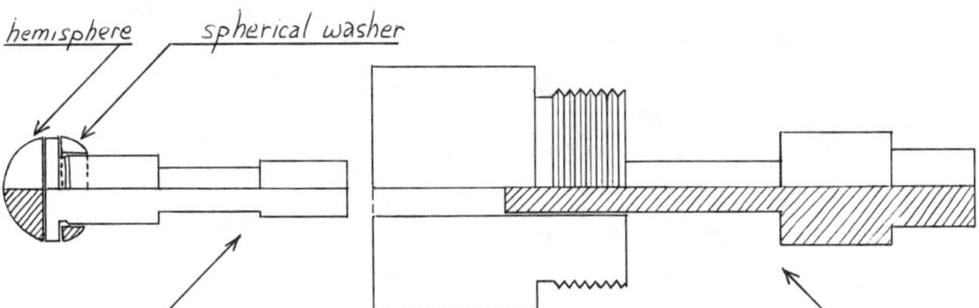

A Mity Mite punch and die set.

with a die. It is very slight and unimportant for reloading, but in bullet making we are dealing with vastly smaller tolerances. The Mity Mite has its operating handle balanced between the links on the ram centerline. The links are jig-made in pairs, and the entire press is machined and fitted, even to the honing of the ram bore, so that accuracy and full use of power are attained. Also, with half the stroke the Mity Mite is automatically twice as powerful as any four-inch-stroke loading press.

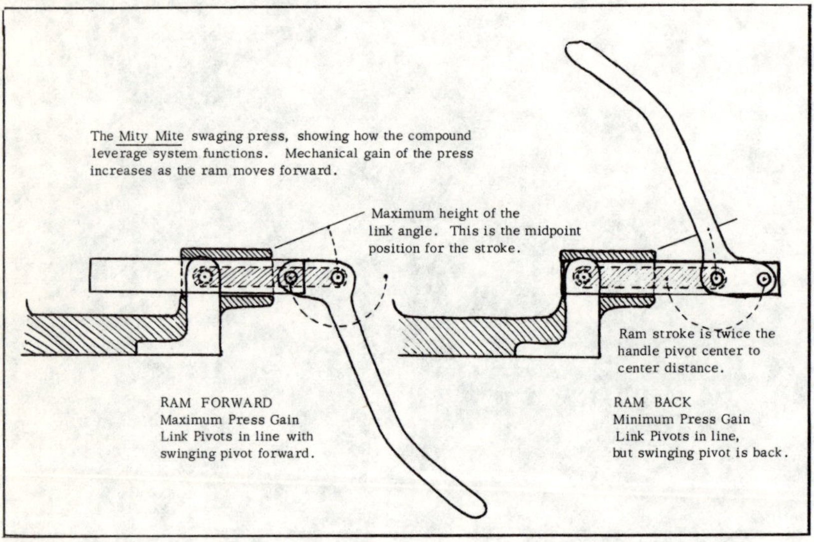

Mity Mite compound leverage system.

One other important feature for the commercial or high-volume custom bullet maker is the automatic ejection of the Mity Mite. In our other descriptions of bullet swaging dies and punches, we have talked about using an adapter and a punch captive within it (the top punch), and tapping this top punch to eject the bullet. Even though automatic ejector frames can be made to clamp onto the ram and pass over the top of the die in a regular loading press, they are cumbersome and not entirely satisfactory for big calibers. We do produce a compound leverage power ejector, which operates like a loading press with reversed stroke, and slips over the top of our dies for easier ejection.

But for simplicity and speed of operation, ejection by means of holding the die in the threaded ram top, while a pin passing through the frame of the press contacts the back of the ejection punch as the press handle is pulled back, is the very best way. The bullets pop out with no further work than merely pulling back the handle of the press. It would be impossible to write about bullet swaging presses without covering this press in at least some detail since it remains as the last of the good-quality short stroke presses on the market.

One other press of extremely good quality is the Hollywood I-type press. Hollywood swage dies operate in a different manner from the dies we have discussed. They are similar to the concept of the old R.C.B.S. dies, in that a bullet is made by pressing the jacket and core into a two-piece die, rather than two separate dies. This die has the cylinder portion made in one part, and the nose portion in another. They fit together very precisely, and to get the bullet out, the bullet is pushed out of the cylinder portion after taking the die apart. Production speed is usually quite slow on this kind of die, and pressure required is much greater. But it has the advantage of being able to make lead-tip bullets in one step. The two-die sets are better suited to open-tip bullets.

A press with compound leverage can generate an awesome amount of pressure. Most bullet swagers don't realize the potential of their press in terms of sheer pounds per square inch. Certainly most reloaders don't ever consider how much sheer force their press can apply on the end of a small punch.

The leverage system of a compound press is much like the leverage exerted by a Luger locking system, in that it depends on a continual change in the ratio of the short end of a lever which eventually approaches zero. A Luger is locked by a toggle joint. When the toggle is closed, it forms a straight line between three pivot points and is in effect a solid bar. Pressure applied in the same straight line would have to shear off a pin in order to open the action. But breaking the "knee" joint just a tiny bit will give the entire system a little leverage, and soon it comes open with less and less force as the knee joint continues to fold outward.

The press uses a similar idea. It is merely a lever with a pivot and two lever arms. One is attached to the ram, and the other is operated like a handle. The pivot, instead of being fixed to a point on the frame, is at the end of two links. The other end of the links fastens to the frame.

Now, instead of having a fixed power multiplication factor of the big lever arm divided by the little one, as with a conventional press, the compound leverage press can have a fixed length of lever for the handle, but an ever decreasing effective lever for the little section attached to the ram. This is because the effective leverage is merely the distance from the ram pivot to where a fixed frame pivot would normally be to get the same stroke at a given point in the ram travel.

The pivot at the end of the links, which is in reality the pivot between the handle of the press and the short part of the lever attached to the ram, continually swings as the handle is moved. When the ram goes forward, the pivot gets closer to the centerline of the ram. As this happens, the effective leverage, which can also be defined as the handle travel divided by ram travel, gets greater at a rapidly increasing rate. Finally, when the moving pivot lines up with the ram pivot on the centerline of the ram, there is no effective ram travel, no effective distance between the two pivots for leverage, and thus infinite power.

How can we use infinite power if there is zero travel? We can only use it to sustain a pressure. But the real greatness of this system for

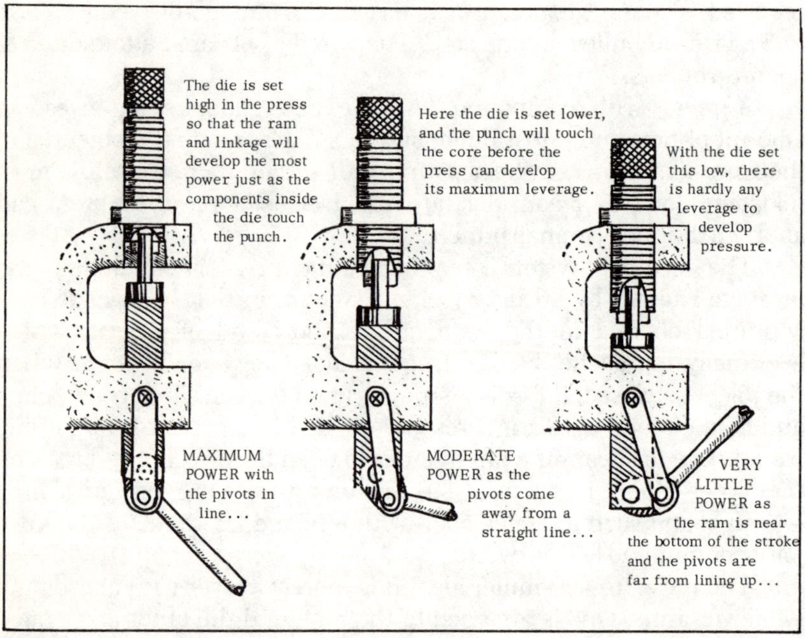

Setting a die for proper press gain.

bullet swaging is that it moves the bullet quickly into the die and then, just as the bullet starts to form and requires greater and greater pressure, the leverage changes smoothly to provide it. At the last instant before the pivots line up, the ram is barely moving. This is the point at which we can put almost the complete theory-provided power to use—and it is enough to wreck the die, punch, or the ram. The only thing that saves the normal reloading press from cracking or breaking a pin during bottomed-out condition, with the pivots lined up and almost limitless power available, is that the entire press has pin and hole tolerances, soft metals, and spring that takes up what little movement is possible at the extreme power end of the stroke.

In the swaging press, we adjust the die so that we can make use of all the available power. In doing so, we have to make sure that we don't create a condition dangerous to the die and punch set. A very little pressure on the handle can create multiplied force as great as 100 times more than applied. If a 150-pound man were to press just half his weight on the press handle while the press was set to contact something solid near the end of the stroke, in the vicinity of the 100X power factor, it would be possible to apply 7,500 pounds of force to the ram and whatever the ram was contacting.

If the ram happened to be fitted with a .224 caliber rod, pressing against a core or a bullet in a die, then all this force would be passed through an area of 0.0394 square inch. That amounts to a pressure of over 190,000 pounds per square inch! Long before this, the punch would have shortened, the die swollen to relieve the pressure, the ram top folded in, or the pressure been relieved in some other damaging way. We start to worry about a nice big rifle chamber with perhaps an inch or more of steel at the breech, when pressure rises about 60,000 psi for just a few milliseconds. The press can sustain pressure in the little die all day if we want to hold it.

The ratio on a normal, single lever press is the handle length divided by the distance from the pivot to the ram at any length. The handle might be 18 inches long, and the short lever about 1¾ inches. Here we'd have a power multiplication factor of perhaps 10.3X. If that same fellow got irritated and threw his whole weight of 150 pounds on that little press handle, as some folks are apt to do, he could generate 1,545 pounds at the ram. Again, with a .224 punch this means a pressure of 39,200 pounds per square inch in our little die! And not just for a few milliseconds, either, but for as long as it takes the swager to realize he's just ruined another die.

Our good friend Ted Smith of S.A.S. Dies once wrote in his old book of swaging that the instructions for operating a die ought to be "put the die in the press, put a core in the die, now throw all your weight on the handle. Take out the die you just ruined and send it back with a check for a new one." At least, that would be a way to keep diemakers busy, not that they need any more diversions.

The loading press and the swaging press both generate far and above any reasonable pressures for swaging bullets which ought to be swaged in them. It takes reasonable care and understanding to use a press for swaging, and a small degree of experience with the proper "feel" of the press handle. We can actually feel the lead expand in the core-seating operation, and we can feel the point forming in the forming die. We can feel it when a jacket folds, and when we have broken a die we can certainly feel and hear that. What is hard to feel is sorry for those unfortunates who get started without any preparation, and commit the cardinal sins of swelling, crushing, and cracking. It isn't hard to avoid, once we know what to expect.

The use of "cheater" bars to extend the press handle for more leverage is, of course, frowned upon. Sometimes it is frowned upon with particular expressions of displeasure, if a custom die worth more loving hours of work than can ever be received in payment is the victim of the proceedings. The makers of presses have enough trouble with this sort of thing in regular reloading. The idea that swaging requires complete reversion to animallike application of brute strength, rather than any sort of humanlike application of control, is probably responsible for the bulk of the damage done to presses and dies. There is certainly a need for a good deal of pressure in swaging a lead slug and a copper alloy jacket to bullet shape—a good deal more than anyone would normally use in reloading. But the pressure is there simply by virtue of a small area and adjustment so that the work is being done at the most powerful part of the stroke. One need not hire a gorilla to operate the handle. Any child can do it.

Speaking of children, it is best not to leave the punch or the die in the press if wee ones are likely to try their hand at operating the handle. In the first place, the punch which slips up inside a small hole in a die is ideal for pinching off or punching through a finger. And in the second, if the punch gets jammed into the die face, one or the other will be destroyed.

Children, and sometimes curious visitors who ought to know better, are likely to want to know what that funny-looking thing is in Daddy's loading press. Upon being told, or sometimes shortly before, their impulse is to grasp the press handle and rapidly work it up and down a few times, to see if anything happens. Chances are, it will. The punch will slip over just enough to strike the die instead of going into the cavity, and Daddy can explain that he *did* have a new swaging punch and die set to make bullets, but now he just has a die.

The safest thing is to remove the punch whenever the swage set isn't being used. Either that or lock the loading room, chain down the press handle, and hope. If the punch that forms the bottom of a rifle bullet gets just a little dent in the edge, then it will start making bullets with a little extrusion that matches that dent, right on the edge of the base. That doesn't help the accuracy. And a handgun nose punch that gets bent a little on the thin edge of the cavity starts making rather lopsided bullet noses with lead extruded in a stream alongside the nose, like an antenna on a police car.

One of the first things a bullet swager expresses at finding out how powerful his press really is, normally, is a curiosity about when it will blow up and kill him. Actually, the pressure in a swaging die may be quite high compared to the average pressure of a firearm chamber over the same time period, but any slight relief will remove the total pressure. There is no gas in the die being compressed, or trying to escape. The energy contained in the pressure is very slight. The energy contained in the gas pressure of a firearm chamber is vastly greater.

When a swaging die fails by cracking, it makes a more or less sharp crack and all the pressure is gone. No bits and pieces fly around the room. But more than likely the die will simply swell that 0.0002 inch or so it needs to move away from the pressure of the ram. Unfortunately, it only swells in the middle where the pressure is, not all the way to the mouth. This creates a jug-shaped cavity. The lead fills the jug. We try to press it out and it is bigger than the mouth of the die. And we either have a hard-ejecting die or one that is permanently fitted with a bullet.

A person might think that this is mostly theory, since firearms can be blown up without the chamber swelling too much. But the cartridge in a chamber is springy, much more so than the lead-backed jacket. It will release even if the chamber is swollen somewhat. The swaging die barely even ejects if a scratch or reamer mark smaller than those acceptable in a chamber is present. A swelling of only a few ten thousandths of an inch is too much for the swage die. And that's mighty small, like slicing a piece of paper forty times edgewise and picking out three or four of these invisible sheets to represent our amount of swelling. Go ahead, try it!

We have sectioned any number of dies with hard ejection and found visible swelling in the die. To be visible, it almost has to be a thousandth of an inch or more. A ten thousandth of an inch is ten times smaller. A human hair, brown, put in a fine micrometer measures about two-thousandths of an inch. Compared to these kinds of tolerances, even a political promise seems substantial.

Chapter 7

Pressure Problems In Swaging

Bullet swaging depends on the use of tremendous pressures—often as much as the average pressure with a gun. The small physical size of the dies and our familiarity with the normal use of a loading press may lull us into thinking that such pressures are not possible with home swaging. But there are two strongly dissimilar characteristics of the pressure occurring during swaging and that occurring in a firearm chamber that tend to hide the truly awesome nature of the force within the swage die.

First, the pressure applied in the swage die is not from compressed or heated gas. It will be released completely in the very smallest movement of any of the components that are containing it. A swaging die with a pressure of 5,000 pounds per square inch within it will not blow to bits and make a loud bang if it should fail because as soon as the walls of the die move back even a few ten thousandths of an inch, the pressure may be completely relieved. There is very little stored energy in the pressure sustained within the die.

The chamber of a gun with 5,000 psi being developed is holding back the same pressure but considerably more energy. The pressure will not be relieved by a mere fraction of an inch of movement of the steel walls; it will not be completely relieved until the bullet is blown from the barrel and all the gas escapes. The gun chamber may even sustain pressures as high as 50,000 or 60,000 pounds per square inch without damage, because this is only a rapid, transient pressure that is gone within a few milliseconds. The very inertia of the chamber walls helps resist such a fast pressure peak. The average pressure over a longer time, from zero to peak and back to zero again, will of course be much lower.

But the second difference between firearms pressures and swaging pressures is evidenced by the fact that the peak and average in swaging pressure may very well be the same. The pressure is maintained as long as the handloader keeps pressing on the press handle. There is no rapid peak to be held by inertia of the die walls, and the tiny die must sustain all the pressure for as long as the handloader wishes to apply it, without swelling in the slightest.

If we assume that a loading press can multiply the force applied to its lever by a factor of about 100, then we might assume a normal man could apply three-quarters of his weight to the press handle without too much strain. Let us suppose that this is only 100 pounds. This means that the loading press will apply a force of 10,000 pounds to the ram. This would represent a force of five tons. Now, if the ram is pressing on a punch, and the punch is only .224 caliber, then it is applying this force over a very small area. The area of the punch end is only 0.0394 square inch. The pressure would be a massive 254,000 pounds per square inch!

More likely, a press might be operated so that it obtains about a half a ton of force on the ram. This results in a .224 caliber punch having a mere 25,400 psi applied to its end, and that would be the pressure within the die. Larger calibers would have less pressure for the same amount of strain by the operator, as the punch diameters are larger. But the very small calibers such as 6mm, .224, and small pistol calibers like the .25 ACP mean that it is rather easy to develop far more pressure than any swaging die could possibly handle. Not only can the die be swollen and ruined, but the punches can be mashed like a tent peg, and the rams of the loading press can be shortened and warped under this kind of abuse.

The ordinary loading press can, without too much difficulty, press a quarter-inch pin right through a quarter-inch-thick piece of cold rolled steel if the two parts are guided properly. It may not survive the experience with any accuracy in the bore to press head alignment left, but the power it can demonstrate is amazing. The special swaging presses, such as the Mity Mite horizontal press, with their two-inch strokes and center-fed compound leverage systems, can develop enough power to destroy any sort of material placed between the ram and the press head as the press passes over center, limited only by the self-destruction of the press.

To understand better about the pressure development in a loading press, let's consider the leverage system of the Mity Mite. This press, first introduced in the 1950s, has a short stroke designed to handle bullet swaging with maximum efficiency but is similar to the compound leverage presses used for reloading. The lever arrangement is such that the ram moves rapidly from its maximum "back" position away from the press head, but for the same number of degrees of handle swing the ram moves increasingly more slowly as it approaches the head. Finally, at one point, the ram stops moving. This is what we might call "top of the stroke" or "top dead center" (referring to terms used in automobile timing). The handle of the press is still moving as the ram approaches zero forward movement. The power of the press, or its multiplication of the force at the handle, is the ratio of handle movement divided by ram movement.

Thus, as the ram movement approaches zero, the quotient of handle movement divided by ram movement approaches infinity. This is why the press is limited in maximum power primarily by self-destruction. If the ram is blocked just a quarter-inch short of "top of stroke," the leverage is still fairly low in ratio of handle to ram travel. It would take considerable effort by the handloader to hurt the press at this point. But place a solid block of steel between the press head and the ram a little further up in the stroke, and the force begins to build without much effort by the operator. If we cut a piece of steel just thick enough so that it contacted the press ram a few thousandths of an inch before the top of the stroke, then a person could easily, with one hand, develop enough power to destroy the press.

This is the key to all swaging and drawing operations in a compound leverage press. The change in available power is slowly increasing during the greatest part of the press stroke. But as the ram nears the top of its stroke, the power it can produce with relatively little effort by the handloader increases very rapidly. Some presses may not develop the power as rapidly as the short stroke Mity Mite, but any compound leverage press will develop more than enough power to swage jacketed bullets. Adjusting the die so that neither too much force is applied nor too much effort required is the heart of successful home swaging.

In resizing cases or seating bullets in a loading press, one of the common procedures is to adjust the die so that when the ram is raised all the way to the top, the face of the sizing or seating die just touches the ram top, or the top of the shell holder on removable shell holder rams. Then the internal parts of the die are adjusted for bullet-seating depth or for knocking out the primer. This is never done in bullet swaging by the experienced handloader. The first step is always to start with the die adjusted just barely into the press threads.

A core or bullet is put into the die, or placed over the ram's punch, and the ram is carefully raised. Normally the core or bullet will not touch the die even with the ram all the way up. The die is then screwed down gently by hand until it touches the core or bullet. The ram is backed down slightly, and the die given a very small adjustment downward. This might only be a quarter turn. Then the ram is raised, and this procedure repeated until the core or bullet begins to be formed as desired.

In the core-swaging operation, which is the first step in making an accurate rifle bullet, the core swage is never used to adjust the core weight from a gross overweight core to a desired core. The cores are cut only very slightly longer than desired. Or, if cast cores are used, the mold is adjusted so that cores only a few tenths or in no case more than about three grains heavier than desired are formed. The core swage is operated under rather high pressures, as it must actually extrude any difference between the final core weight and the raw cores put into it. The ram is adjusted so it passes over center (top of the stroke) as the desired weight is formed. In doing so, almost no limit is placed on the pressure that can be generated; this is controlled by the bleed hole sizes in the core swage die and the amount of lead that has to be bled off, as well as the hardness of the lead.

Core swages can be split like a ripe melon from the use of hard lead alloys or from forcing more than a small amount of lead through the bleed holes as the press goes over the top. The only amount of lead that is bled off is the amount of extreme variation from the heaviest core to the lightest. If this is only a grain or two, then relatively little strain is placed on the die. But because this die must have the smallest punches and bore for the entire set for a given caliber, it will take the greatest amount of abuse from excess pressure. Unfortunately, it is the first die the handloader is likely to use and the easiest to destroy.

It is tempting to use the core swage to adjust core weights, rather than to simply eliminate the small differences in similar length cores. But if this must be done, it is best to take it in very small steps; running all the cores through in two separate operations is better than taking one big step in each core. This allows the extrusions to be cut off in short sections as the cores are ejected, rather than forming long strings which must be moved out of the holes as a single unit. Lubrication is improved since lube is used on every core, and the lube goes between these short sections, helping reduce the friction between them and their extrusion holes.

The core-seating die is the next die which is easily damaged by excess force. It is a sealed chamber, with two punches fitting tightly on either end and no escape for pressure. At least the core swage has some relief from extrusion of lead, while the point-forming die has an open tip. This die is normally adjusted down so that it is a little lower than the maximum power point when the core is being seated. This permits the handloader to feel the force he is applying better than if the ram passes over center. The right amount of force is just the amount necessary to seat the core snugly, bringing the jacket to the die walls and a faint rim of lead up to the junction of the bottom punch and the jacket's inside wall.

An undersized core-seating punch lets the pressure escape, and a very small amount of tolerance is necessary for air to escape from the core and jacket. Too small a punch simply extrudes lead around it. But a slightly undersized punch isn't too bad for a beginner to use, as it still makes good bullets and at least gives some chance of saving the die should he become too enthusiastic about seating the core. Lead extruding around a punch which is only about a thousandth of an inch smaller than the inside of the jacket usually indicates too much force. Benchrest shooters usually use more force than a shooter who is making hunting bullets or handgun bullets needs. But even this is done with great care and moderation.

The amount of damage that is done to a die is most often not visible. It usually consists of swelling the middle of the die only a few ten thousandths of an inch larger than the mouth. Sometimes the die will swell as much as two- or three-thousandths of an inch, like the overheated beer can. These amounts are hard to see, but they can be felt in ejecting the core or bullet. The swollen die has to resize the core or bullet as it is being driven past the mouth, taking much more pressure than it normally would to eject. This places a terrific strain on the top punch and ejecting system, and can warp or gall the punch against the die wall. A bent punch then will not slide up and down smoothly, and sometimes prevents the insertion of another core or bullet.

In the point-forming die, the pressure is relieved somewhat by the open end of the die. Too much pressure here normally is a result of forcing the bullet too far into the die. The core can extrude up the tiny ejection pin hole, so that when it is ejected the bullet will have a button with a hole in it at the tip. This is the folded pipe of lead that squirted up the ejection hole. If the jacket is forced up into the ejection hole, it will sometimes require enough force to make the bullet stick firmly in the die. Point-forming adjustments are usually done with a lead slug of the same weight as the bullet, well lubricated, and formed up just short of the ejection pin-hole. The same slug can be turned end for end and continually reformed as the die is adjusted for a nice bullet shape.

The final adjustment to the point-forming die is done with the jacketed bullet, and is usually very close to the point at which the same weight of lead slug formed up properly to a point. Generally this work is done as the press is adjusted to pass over the top, for maximum ease of forming and uniform tips. An ejector frame that attaches to the ram, or one of the cam or lever types that slips over the top of the die, is advisable for the most uniform points. Ejection with a mallet usually means that the point will not be quite as even as with a mechanical system. Presses such as the Mity Mite have automatic ejection features on the press backstroke.

In all steps of swaging with jacketed bullets, the jacket should be smaller than the die into which it is to be placed. This difference need only be a thousandth or even as little as a few ten thousandths. But if the jacket is the same size or larger than the hole in the die, it will normally be hard to swage, harder to eject, and probably stick. The amount of spring in the jacket is responsible for this. The jacket needs to spring back just a tiny amount from the die walls, similar to a case in a rifle action, in order for it to release its tight grip more easily. A jacket already too large for the die will spring back toward the die wall, making ejection very hard. Imagine trying to chamber a cartridge too large for the rifle! This is the same effect as swaging with an oversize jacket.

Jackets can be drawn down in special jacket-drawing dies, if the proper size is unavailable for a certain caliber. In some cases, a jacket may be the entire cause of many problems that appear to be caused by excess pressure or a bad die. An undersized jacket makes a more accurate bullet, but only if it is within a few thousandths of finished diameter before core seating. Using a .44 caliber jacket in a .45 swage die even works reasonably well, with perhaps a rounded bottom edge to the bullet as the only unusual result. But using a .38 jacket for a .44 bullet is going too far; it will usually form up only with great pressure and then will have lead extruded all around the jacket mouth in a very uneven manner. In an emergency it can be done, but the results are less than satisfactory.

Extremely high pressures can be generated when reswaging jacketed bullets or hard lead bullets to another shape or caliber. The practice should only be attempted by handloaders who are completely familiar with normal swaging work and have experience in the more advanced uses of swaging. In some cases very good bullets can be formed from otherwise poor shooters by using this technique, but the difference in diameters or shapes should not be too great. We once made a swaging die to re-form .223 caliber military bullets to nearly the same spitzer shape they originally had, but with a more evenly formed base. The open base of these bullets had rather extreme variations in the amount of depth before lead was encountered, and had uneven distribution of lead at the base. A reswage die was made for a customer who had several thousand of these bullets and had very poor groups with them. They reswaged rather easily and in the process developed a more even base with resultant tightening of groups. But the difference in diameter was only a tiny amount and the die was made slightly larger than the original bullet.

Pressure is the tool of the bullet swage, its fuel and its essence. But the handloader must pretend to be a hydraulic press in his operation of the mechanical one: slow and powerful, under complete control at every moment. Jerking the press ram up and down or throwing our weight on the handle are sure ways to ruin the die, the press, or the punches. Swaging is a precision application of tremendous forces which must be kept under control. The press will do as it is commanded, either correctly or toward self-destruction. If there is one point about swaging which must be completely understood, it is how much power is generated when the press passes over the top of its stroke, and how much damage can be done by careless operation.

Chapter 8

Swaging Lubricant

Lubrication of the core prior to swaging, and of the jacket prior to seating or point forming, is necessary so that the high pressures will not make the lead or jacket material bind itself to the die walls. The lubricant has to be able to maintain a very tough film under extreme pressure, yet not pile up or push ahead of the forming material and leave bare metal-to-metal contact exposed. Most of the lubricants used for bullet swaging are also usable for case forming and reloading, but the reverse is definitely not true.

There are a number of suitable lubricants, including some of the motor oil additives like STP, but all of the heavy synthetics seem to have the problem of being messy to handle and hard to remove from hands, clothing, and bullets. Two of the very best lubricants are anhydrous lanolin (an animal oil, from sheep) and a special drawing lubricant manufactured just for the purpose, called Draw Die Lube. Another excellent swaging lubricant is a mixture of half lanolin and half castor oil, melted together to form a thick paste. Because the lanolin base lubricants are corrosion inhibitors if pure ingredients are used, they are best for the handloader who is not particularly careful about cleaning up the dies and bullets after working with them. The disadvantage is that lanolin-based lubes have different consistency with temperature, melting to a thin fluid at summer temperature levels and becoming a hard paste at winter levels.

The synthetic Draw Die Lube is water-soluble, as is lanolin. Cleanup is very simple with either lubricant. Simply wipe out the dies with hot water and dry them with a cotton swab, then oil the bore and punches. Bullets can be put into a pan of hot water, and dried from their own heat. In all the steps of swaging, it is assumed that the outside of the core, or jacket, has a film of lubricant applied to it. The best way to apply a film of lubricant quickly is to moisten the fingertips with it, and roll the bullet, core, or jacket between the thumb and forefinger before putting it into the die. Lubricant must not be allowed to get inside the jacket; if it does, it must be washed out. Any lubricant inside the jacket will be trapped by the core and throw the bullet off balance. The lubricant keeps the core from bonding to the jacket, just as it keeps the jacket from bonding to the die.

For very precise work, the amount of lubricant on the jacket will affect the diameter of the bullet. Rather lengthy procedures are sometimes used to ensure that benchrest bullets get exactly the same even quantity of lubricant film applied. One of these is to saturate a small cloth pad with lubricant, and then lay another cloth over it. A roller is used to press the two together. The second cloth is then laid on a clean stamp pad and rolled against it. By this time, the lubricant is spread very evenly and simply rolling a bullet across the stamp pad will put a very even film on the surface. Another way is to measure out an exact amount of lube such as 2 cc, from a graduated beaker or a hypodermic syringe, putting it in an even line at one end of the pad. A spatula is used to spread this amount evenly across the pad. The quantity is adjusted so that one pass will just cover the pad.

Actually, for most practical work rolling the bullet between moistened fingers will do the job just as well. The effect is so small only a benchrest shooter would want to worry about it. The effect of too much lubricant is the same as too much case lube when resizing brass: dents occur in the ogive. These actually take the appearance of scratches. Some very similar marks occur from the natural folding of the jacket material, especially with very thick jackets folded to make a very small tip. If factory bullets are inspected closely, the same kind of mark will be seen on them. The appearance is made more pleasing by tumbling or vibrating the bullets with a polishing compound, such as walnut shells and rouge.

The small lube "scratches" are not harmful to accuracy. Probably tumbling them is more so. Yet, since part of the shooting game is believing that something very shiny and bright has to be more accurate than something with marks or a dull finish, the handloader may wish to tumble his bullets to give them the same factory appearance. Generally, they may shoot just as well either way. It is better to have some marks on the ogive, if they are small, than to have a bullet stick in the ogive die. Too little lubricant will almost always result in a stuck bullet, as will the use of the wrong kind of lubricant.

Because the core swage die not only has the smallest diameter of all the dies in a given set and thus generates the highest pressure under proper use, but also has to work with lead directly in contact with the die bore, proper lubrication is important. The cores should be lubricated so that the force required to eject them is not excessive. Also, some lubrication should be able to extrude into the bleed holes along with the lead, to help keep the pressure down and to act as a buffer between the little pieces of lead being extruded. In this way, they will not form a solid continuous wire but will form only a gently attached series of pieces. The strings of extruded lead are more easily broken apart by the moving ram and/or the fingers, so they don't get in the way and foul the press.

After swaging the cores, they must be washed as soon as possible for two reasons. First, one does not want to accidentally make bullets with lubrication inside them. Second, washing the cores in hot water and then setting them up to dry quickly will remove the lubricant from the pores and help expedite the escape of any trapped air in the pores, which might otherwise seep out after the core was seated and remain inside the jacket. This second reason is also why swaged cores are sometimes left to "rest" overnight before precision bullets are swaged.

To prevent oxidation or the formation of carbonate salts on the lead core, the cores should be stored in air-tight jars or used soon after swaging. If a lanolin and castor oil lube is used, it can be left on cores which are not being used for some time. Draw Die Lube should be washed off right away, since it traps moisture and leads to a sticky corrosion substance forming. Penetrating anticorrosion compounds like WD-40 should not be used on cores, as they will be almost impossible to remove completely in later steps. If the cores are warmed slightly in an oven before being put into an air-tight jar, their heat will expand the air in the jar before the lid is put on, and then create a vacuum after it is sealed up and cool.

The jacket can be lubricated by handling for core seating, and then lubricated by handling again for point forming. Very little is required, only a light film. The amount that remains after core seating is usually enough except for some long spitzer shapes or for making lead-tip or hollow-point bullets. The demands for easy ejection are greatest for the lead-tip bullet because the ejection pin has to press against the soft lead, rather than the hard jacket ends. A little extra lubrication will help keep the ejection pin from pressing into the lead.

In forming handgun bullets in a straight-wall die, with a large lead nose formed entirely by a cavity in a punch, it can happen that the grip of the punch cavity on the lead nose is greater than the grip of the die cavity on the jacket walls. When this happens (such as with a very long or deep cavity punch, or one having rather parallel walls), the bullet will tend to pull out of the die and stay on the punch. Usually just a little more pressure will expand the bullet until it grips the die harder and remains in the die. Then it can be ejected normally. If the pressure to do this seems too much, and the bullets keep sticking their noses in the punch, which can mean that they will be damaged by pulling them out of the punch a little off center, then the jacket should be wiped dry of lubrication (some will remain in the die) before swaging again.

If the problem continues, no lubrication should be used on the jackets, but a film of lube should be put into the punch cavity directly with a cotton swab. Most of the handgun nose punches have a small bleed hole that is actually a blind hole, drilled exactly in the center of the punch from the top. This is to allow for a little surplus core and help adjust the weight of the bullets so they are all the same, using either feel or a positive stop. A small excess amount of lead will form a little wire extrusion into this hole, which is then clipped off the finished bullet. If the hole is filled with lubricant, it will help act as a cylinder and trap a little air, which then pushes on the core extrusion and helps eject the bullet from the punch cavity.

Excess lube in the punch cavity will result in partially unformed lead tips even with normal pressure; so the amount of lube put into the punch cavity should be regulated between the extremes of a stuck bullet nose and a partly unformed bullet nose. Lubricant will remain in the die, also, after each step, so that a person can run several bullets through the die in a row without lubrication, before one finally sticks. This can lead to confusion over why the bullet stuck, since the previous ones might not have stuck and had no lubricant either.

Lubricant can be used to remove a stuck bullet in most cases, by forming a sort of hydraulic wedge to compress the jacket away from the die walls. This is covered in chapter 27. Basically, a short lead core is inserted behind the stuck bullet, after having been actually dipped in lubricant, and the lead core seals the bore of the die completely. The trapped lubricant cannot force its way past the lead; so it seeps between the jacket and the die walls, opening a crack which fills with high-pressure lubricant and compresses the jacket. The jacket can be ejected along with the lead core and lubricant in a normal manner after a little of this pressure is applied.

Water, preferably hot, is recommended to remove lubricant from dies and bullets, but some people feel reluctant to use it and yet don't know what other substitute to try. Water is associated with rain, with rust, with corrosion, and in general with carelessness; one does not want to get fine equipment wet, just from long-standing tradition. But hot water is actually one of the best and safest cleaning substances for bullets and dies, if it is dried out immediately both from the heat of the parts or bullets and from circulating air. It leaves no film, evaporates completely, and is not by itself harmful unless it remains in contact with the material and with the air for a long time. Dies should not be dropped into hot water since quite often the top punch is kept in the die all the time to ensure a good fit and protect the edges, but hot water on a swab can be used to wipe out the die bore, followed by a dry swab to remove the water, and a good oil.

Hot water does not harm bullets in the slightest. It removes the recommended lubricants quite well, and the heat absorbed by the bullets dries them out quickly if they are spread out and air is circulated over them. A warm oven, not over 250°F to 300°F, can be used to hasten the drying. An alcohol without water or glycerine added (not rubbing alcohol, which has these in it) will also absorb the water and will dry without leaving a film. Solvents can be used, but unfortunately they tend to pick up oils and then leave a film on the bullet. Anticorrosion sprays used on finished bullets may affect the powder charge, and are not recommended. One of the inert wax compounds is better, if anything is desired.

Once in a while, a person will confuse bullet lubricant, such as Alox-beeswax or graphite grease, with swaging lubricant, since both are used on bullets. The bullet lubricants for cast bullets are nothing like swaging lube. They will not work at all. Neither will swaging lube help a bit in easing the passage of a cast bullet down the barrel, since swaging lubricants are primarily liquid while bullet lubricants are thick, almost dry greases. It is sometimes suggested that cannelure grooves be rolled into swaged jacketed bullets and regular bullet lubricant be put into them to assist in getting high velocity or to reduce barrel wear. Again, this is a misapplication of the material and only shows a lack of complete understanding of the purpose of the bullet jacket, which is to isolate the lead from the barrel, powder gas, and air friction, as well as to control performance and afford a better grip for the rifling.

Lubricating a jacketed bullet with a lube suitable for cast bullets is not only futile, but probably harmful to accuracy. The high velocity of rotation of most jacketed bullets, which is a result of their being used in the first place instead of cast lead, would throw most if not all the lubricant out of the groove. If it did manage to throw all of it out at once, possibly the bullet would still be balanced. But if any remained longer on one side or the other, the bullet would spin about an unbalanced axis with the center of form. Since the jacket will be impressed considerably by the rifling, usually about four-thousandths of an inch on each groove, it is most likely that the lubricant will be wiped completely out of the groove and will have a minimal effect on friction or heating at the levels it occurs with most jacketed bullets and the loads they are generally used with. The alloy of the jacketed bullet is its own best lubricant.

Another common misconception is that a high-pressure lubricant means one that is applied with a high-pressure grease gun. It may seem foolish to some shooters, but still the question is asked often enough to be worth answering here. Grease applied to machinery and automobiles with a high-pressure gun is not at all the same thing as a lubricant that will sustain a good film under high pressure. Any very heavy grease that remains fairly thick under pressure will probably not be suitable for bullet swaging, since the thinnest film that will still prevent die-to-jacket contact is desired. Most of the thick greases will be pushed away by the leading edge of the bullet or will tend to gather in pockets and dent the jacket.

Still, there are odd lubricants being used every day for swaging, and the handloaders are getting away with it. It isn't recommended, but some people do use shortening grease or oil, some use a favorite brand of hair oil (many contain lanolin), and a few even swear by oddities like chicken fat or bear grease! Each to his own, but for the purposes of warranty there are two lubricants: lanolin and the 50 percent lanolin-castor oil mix, or Draw Die Lube.

Chapter 9

Working with Lead

Lead is the basic element which accounts for the bulk of weight and volume in all bullets. It is essential to understand how lead flows under pressure in order to design swaging equipment, and it is interesting (if not necessary) to have a good background in the physical properties of lead before working with it as a handloader.

Lead is one of the oldest known elements, being mentioned in the Bible and thought by alchemists to be the first metal. It is obtained chiefly from an ore called galena (PbS), and also from anglesite, cerrusite, and other oxides, sulphates, or carbonates. Lead rarely occurs in nature as a free element. Pure lead is bluish white with a high luster when freshly cut. It soon oxidizes or forms carbonates on the surface which are insoluble and protect the rest of the lead from corrosion. Because of this fast-forming, tough film, lead containers are used for many acids. Nitrates and acetates of lead are soluble however. Because of its corrosion resistance, lead was used as water pipes and for fixtures in ancient Rome. Some of the same pipes are in use today. Lead is also used in solder, type metal, bearings, storage batteries, cable covering, and even gasoline. Lead tetraethyl is the material which is being slowly phased out as the newer cars come on the market.

Lead is a cumulative poison. Care should be taken in handling it or when melting it, so the fumes or salts that form on the surface aren't taken into the body. Particularly deadly is lead arsenate, used as an insecticide in more innocent times. White lead (PbSO), yellow lead ($PbCrO_4$), and red lead (Pb_3O_4) are used in paint. Lead is a good vibration and sound absorber, and is also used as a shield for X-ray equipment, radioactive materials, and around some areas in nuclear reactors. It is obvious that there are many uses for lead. The handloader need not worry about the world supply drying up or laws being passed to eliminate lead from availability.

Lead is a dense, easily drawn, and easily melted material. It is relatively cheap compared to most other metals in the same density range. These things make it a natural for bullet swaging or casting. In swaging, ductility is more important than the melting point. The density of pure lead is 11.34 grams per cubic centimeter at 20°C, or 0.4097 lb/in^3 at 68°F. Since there are 7,000 grains to the pound, lead weighs 2,867.9 grains per cubic inch. This is good information for calculating the weight of bullets of various shapes before actually making them.

Lead has the chemical symbol Pb, which stands for the Latin name of lead, *plumbum*. It is the eighty-second element, with an atomic weight of 207.21. Its melting point is 621.3°F$\pm$0.2°. It can absorb 0.031 calorie per gram per degree Centigrade at 20°C, and has a heat of fusion of 11.3 Btu per pound. Lead will expand 16.3 microinches per degree F, from 68°F to 212°F, in a linear direction. Its modulus of elasticity in tension is 2.6×10^6 psi. It is a barely compressible material, which will flow rather than reduce its volume under high pressures.

A common alloy of lead is an industrially pure lead with 99.93 percent lead and 0.06 percent copper. The density of this material is 11.36 gm/cm^3 at 20°C (room temperature), and it melts at 325 to 327°C (617–620°F). The coefficient of thermal expansion is 2.9×10^{-5} per degree Centigrade. Modulus of elasticity is 2 million pounds per square inch. The tensile strength of a cast slug of this technical purity lead is only 1,750 pounds per square inch, whereas an ingot of iron would compare with roughly 40,000 pounds per square inch. The yield point of the lead is 1,000 lb/in^2. Copper, in comparison, has a yield point of 12,000 for soft- and 50,000 for hard-tempered.

In the tensile and yield point testing, it has been found that lead of this standard commercial purity elongates 40 percent in a two-inch section and reduces in diameter by 90 percent. It has been assigned a Brinell number with the 500 kg force of 5.5, which would compare with a range of 21 to 55 for various purities of aluminum on the same Brinell scale. Lead can be made to flow and extrude most easily at temperatures just below its melting point, and the heated ingot is the best way to extrude large rolls of lead wire for cores. Generally the die is pressed into the ingot, rather than the entire mass of lead being forced along with a plunger to the fixed die. This reduces the necessary force to extrude a given size of wire.

Since lead is the primary component of bullets, it determines the bulk of their weight. It is very useful to know the shapes and formulae for volumes of these shapes into which bullets can be broken down for study. The basic part of a bullet is merely a cylinder. The area of a circle is 0.7854 times the diameter squared, so the volume of the cylinder is merely this area times the height. This could be written as $v = 0.7854hd^2$. A ball or sphere has a volume equal to the cube of the diameter times .5236, written $v = .5236d^3$. The volume of a cone is equal to the area of the base times a third of the height.

Using some of these shapes will give combinations for calculating the volume of bullets, and this will enable a handloader to calculate the weight before the bullet is actually made. Some convenient conversions for using various density and volume formulae include:

0.0648 gram = 1 grain = 0.0353 ounce
1 cubic centimeter = 0.061 cubic inch
16.387 cubic centimeters = 1 cubic inch
1000 grams = 35.2738 ounces = 2.2046 pounds
70.308 kilograms/square centimeter = 1000 pounds/square inch

The question is often asked, is there any other metal that would make as dense a bullet and thus would have as excellent an ability to carry energy against the force of air friction as lead? There are several superior metals and alloys with greater density than lead. The difficulty of working or the cost of obtaining them is the main reason lead is used. Some of the metals which would make bullets that would potentially be superior to lead in external ballistics include: iridium, gold, gold and copper alloys, osmium, palladium, platinum, platinum and iridium alloys, rhodium, ruthenium, tungsten, and the heavy man-made elements which do not exist in sufficient quantity in any case.

Lead weighs 0.409 lb/in³, or 708 lbs/ft³, which compares with 0.698 lb/in³ for tungsten. A core of tungsten would be expensive but not beyond reason. It would have to be sheathed in lead or some soft alloy so that the bore would not be ruined by having this hard metal pass through it. Such bullets are available commercially at this time, and offer very fine range and flat trajectory but at a high price per bullet. Since the alloys of lead are considerably less dense than pure lead, the swaged bullet with a jacket protecting the core should offer superior ballistics to the cast bullet of identical shape and weight, or of identical volume and shape, since the swaged bullet will weigh more for the same volume in the same shape.

The difference between alloys and pure lead, regarding density, is charted for tin, bismuth, and antimony, for various common alloys:

Lead	*Antimony*	*Bismuth*	*Tin*	*Lb/In³*
30%	70%			0.2604
37%	63%			0.2662
44%	56%			0.2748
63%	37%			0.2974
83%	17%			0.3449
90%	10%			0.3807
33%		67%		0.3697
50%		50%		0.3796
67%		33%		0.3946
75%		25%		0.4033
83%		17%		0.4062
88%		12%		0.4068
30.5%			69.5%	0.2974
46.7%			53.3%	0.3153
63.7%			36.3%	0.3402
84%			16%	0.3726
87.5%			12.5%	0.3825

Lead cores may be cast or cut from lead wire, but in some cases the jacket used will not be large enough in the internal cavity to accept a standard core diameter, yet the next smaller size will result in too much length. The length of the core on smaller sizes can be great enough before seating or swaging, so that the wire will bend or mushroom rather than go into the die. The entire core must be inside the die before any pressure is developed, both to keep the core from bending and to protect the punch from being harmed by being off center with the die bore.

To solve this problem, larger core wire may be drawn down by passing it through a drawing die, made of either tool steel or even soft plate steel, with a smoothly polished and radiused hole. The end of the wire is sharpened, using a knife or even an old pencil sharpener, pushed through the hole, and then grasped with a pair of pliers. If the reduction is not greater than a few thousandths of an inch at a time, it will be easy to draw the wire through the hole, with the plate clamped in a vise. The wire can be reduced to a size that will fit any jacket thickness in the caliber range.

Individuals who wish to manufacture extruded core wire from billets of lead would be able to do so most easily and cheaply using a large hydraulic jack to apply pressure to a die, which is forced into heated lead within a cylinder. The lead should be heated just below the melting point. A die which will not go soft at temperatures from 600° to 800°F would be necessary for long life. Usually one of the tungsten steel alloys will fill this requirement. Heat will be added to the lead by the friction of drawing, and this will heat the die beyond the temperature of the billet. Calculations for the quantity of heating required can be made from the thermal constants for lead, for a given size of billet.

Extrusion of lead requires moderate pressure, but more important it requires a long stroke for production of reasonable lengths of wire. The combination of long stroke and moderate pressure eliminates the use of low-cost presses and cylinders, and requires a fair investment in machinery for economical production. Generally, it is best to purchase continuous rolls of lead wire of commercial purity, for the use of an individual, smaller shop, or club. Continuous lengths can be extruded by loading material behind the billet after each stroke, even with short-stroke presses, but the time of retracting the die, opening the cylinder, and adding material often consumes more effort than the operation of extrusion itself.

Chapter 10

The Core

Lead cores for bullet swaging must be pure, or at least very soft, lead. Hard alloys just won't flow under safe pressures for the kind of dies that fit regular loading presses or smaller hand presses. The easiest way to get cores for swaging is to buy 25-pound spools of lead in the proper size from a fishing supply store, a swage die firm, or the lead smelters. The sizes are usually in English measure, such as 3/16-inch diameter, ideal for .224 and 6mm bullets. The lead wire must always be small enough to fit inside the jacket.

Lead wire which is far too small for the caliber will result in having too long a core. The long, flexible core will fold several times as it is pushed into the die, making air pockets and generally getting outside of the die mouth before it reaches proper diameter. Core swages are used to bring the diameter of lead wire, or cast cores, to a diameter just right for the jacket being used. The core swage also takes the last bit of variation out of the core weight.

Core swages are very much like the core-seating die described with the two-die system of swaging. The only real difference is that they have one or more bleed holes around their circumference. These tiny holes permit a thin wire of lead to be extruded from the die. The die is adjusted in the press so that as the ram comes up as far as it will go, the lead core is compressed, and a few tenths of a grain, or sometimes as much as three or four grains, is bled off through these extrusion holes. If the press comes up the same each time, then the volume left in the die for lead will be the same. The variation in weight will be bled off through the holes. On the ejection stroke, the tiny wires are cut off by the core itself.

The more excess lead is bled off through the holes, the greater the pressure will be in the die. A certain minimum pressure is required to make the lead fill the die properly, but too much will swell the die and compress the punches, perhaps bending them. It is hard to imagine the tremendous forces at work but they can exceed some rifle chamber pressure averages. It is wasteful to swage away more excess lead than absolutely necessary, but a few tenths of a grain to as much as four grains may be required depending on the particular die. Weighing the cores before and after swaging and taking the mean variation of ten cores will result in a good check on the effectiveness of the process.

Lubrication is required in each swaging step, including core swaging. This prevents leading of the dies and makes ejection forces and stresses much easier. The lubricant must not be allowed to stay on the cores, though, because whatever goes into a jacket stays there. Cores lubed with Draw Die Lube or lanolin can be washed in warm water and dried before using. Other lubricants may require a solvent other than water. Cores are not lubed before putting into the core-seating die, but only a thin film of lube is applied with the fingertips to the outside of the jacket.

Much damage by beginners is done to the core swage die, because it is the first die used, the one with the greatest ratio of pressure to volume, and the one with the smallest punches. Most damage occurs when trying to swage hard lead. The next most common problem is trying to adjust core weights with the core swage instead of just taking care of the variation. Odd lengths of lead wire should not be inserted and then brought to some desired weight with the swage. That is not the purpose. The cores should be brought to a weight which is the desired core weight, plus the maximum deviation noted in making about ten such cores, plus a grain or so for buffer. For example, suppose we want to make a 52-grain bullet. We weigh the jacket. It might be ten grains. Then we need 42 grains of core.

But if we used a core that was cut or cast to weigh about 42 grains, it would still have a plus or minus variation from the cutting process of about a grain (in most castes) or half a grain from casting. This gives us a range of cores from 41 to 43 grains. Of course, this is only an example and any given attempt might result in more or less (probably more) variation than this.

If we cut up ten cores, weigh them, add the weights, and divide by ten, we have the mean weight. Let us say it comes to 42.5 grains. The core swage can take care of all the heavier than normal cores by swaging away the excess. It can't add anything to the light ones. If it is adjusted to swage away all the variation, the cores will all have to be too light to make a 52-grain bullet. The cure is to add the greatest deviation in the too-light direction to our desired core weight. If the lightest of ten cores comes out at 41 grains, then the maximum deviation from the mean is 42.5 less 41, or 1.5 grains. By adding another 1.5 grains we can have enough weight so that all the cores will come out at any weight we want that isn't over the mean. To get enough pressure to ensure good core forming, we might have to add another grain or two.

All the figuring can be eliminated by just adding enough extra core weight to cover any possible variation in the cutting or casting process, and then setting the core swage to remove it all, resulting in bullet cores of the desired weight. It is necessary to find out just how much lighter than average the lightest cores come out in order to adjust for the most efficient use of the die. The amount we add just to keep the pressure up in the die is called the buffer, and it is controlled to a large extent by the temperature and the die bleed hole diameter and length. Usually it is possible to get the final swaged core variation down to a tenth of a grain or less in the smaller calibers.

In core swaging, the length of time that the ram is kept forward determines the length of time pressure is maintained on the core. Since the lead is more like a thick putty than a liquid, it will flow for some time as pressure is maintained on it. Eventually it will flow out until the pressure drops to below the force generated by friction in passing through the bleed holes. If the ram is pulled back before the pressure equals the friction of extrusion, variations in core weight will result that are controlled by the time of the ram being forward.

Since some presses can pass "over center" and thus actually pull the ram back a little, it is important to use exactly the same length of time for the swaging of each core, and to move the press handle the same way, the same distance over center, at the same speed each time. Another way to handle this is just to put a positive stop in the edge of the bench so that the handle hits it. Then hold the ram up, with the handle against the stop, for a few seconds (or up to a couple of minutes if desired), timing the cycle with a clock. This is, of course, for squeezing the last bit of accuracy out of the bullet, and is not a requirement for normal swaging work.

Actually the weight of the bullet has less effect on the point of impact than the variations in powder charge, the variations in diameter, perhaps the breathing of the shooter. In benchrest work these things can be controlled, while in hunting just the estimation of distance is probably a much greater factor than any of the other variations. In bullet swaging, where one becomes accustomed to working in ten-thousandths of an inch and in minute amounts of variation, it is too easy to lose sight of the fact that even the barrel of the gun probably will have tight and loose places greater than our level of control in the bullet. One can get too involved in precision for the application, but it is nice to know there is the possibility of control.

Cores which are cut from lead wire need to be cut in a machine that does not bend them, has a good shearing action, and is reasonably accurate in maintaining the same weight from cut to cut. A core cutter ought to have a good thickness of metal to support the uncut wire from bending, and a firm positive pivot without undue slop between the two parts. Commercial cutters using multiple holes are available, so that one cutter will handle all sizes of wire. The hole to use is the one that just lets the wire slip through easily, even if it is slightly bent from handling.

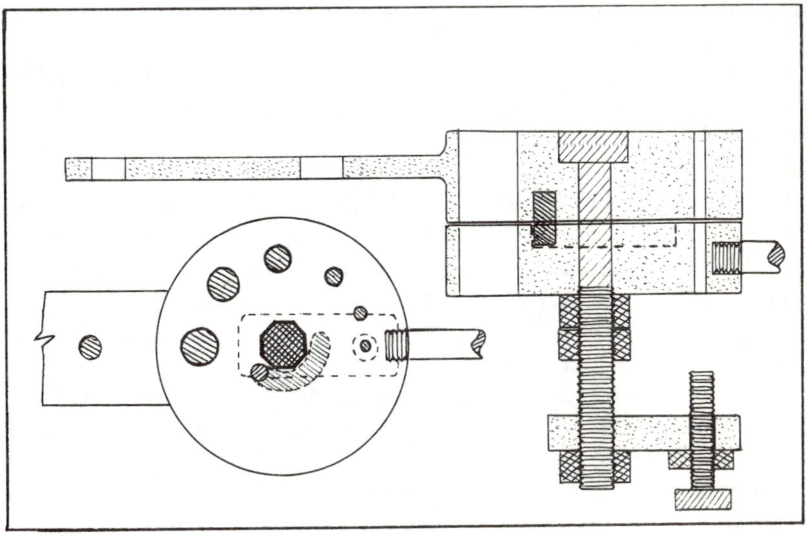

A universal core cutter design of good quality.

A length of wire long enough to stand under its own strength, perhaps two feet long, should be cut from the spool. Normally it is best to wipe the wire down with a clean cloth to remove grit and dust. Sometimes, to polish the wire and give it a better diameter for a given caliber, it is drawn through a simple drawing die by hand. These can be made on special order by diemakers, or they can be made at home. A quarter-inch-thick piece of steel with a gently tapered, polished hole the right diameter for the finished wire is all that is required. The end of the wire is sharpened (an old pencil sharpener works okay) and the end pushed through the hole. A pair of pliers is used to draw the wire through. Green soap or Draw Die Lube on the wire will make it draw more easily.

Lead pipe and telephone cable sheathing are two good sources of soft lead. Linotype and wheelweight are useless for swaging; they should be saved for good hard cast bullets. The idea of swaging is not to depend on the hardness of the lead for preventing leading, but to use the jacket and get the full benefit of soft, dense lead. There is no benefit to a hard alloy other than to prevent leading and gain penetration. It will not expand better than pure lead, or hold together better. If penetration is desired, a double-jacket bullet with the smaller caliber jacket turned over the mouth, inside the larger one, can be used, or any number of other advanced techniques. Suffice it to say that swaged bullets can be made to do exactly what the shooter wants, and no hard alloy is required. More dies are ruined by trying to swage hard lead than any other way.

Cast cores using low-cost lead are a very popular thing these days. Fast, bench-mounted multicavity core molds with adjustable weights are available in any caliber. An open-top lead pot is best for these molds, using a fairly large dipper. This lead should not be fluxed, and the top portion can be skimmed to remove suspected lighter metals (which are also harder). The mold should be rather touchy on one or two cavities as a general rule until it gets warm, and if it is too difficult to get four good casts each time then either the technique of pouring is at fault or the mold needs to be mounted at a slight angle, to let air escape better. A core under the mounting handle at one end will tip it just about enough to do this.

Single- and multiple-cavity fixed molds are also available from time to time, and these are convenient as well as low in cost. But they still require the cores to be cut to weight. Their advantage is lower cost, and elimination of the need for lead wire. For the best accuracy, even with swaged cores, a core from each cavity should be taken individually and all four put into separate boxes. This ensures that even the slight variations between cavities will not be present before swaging. The core swage is a device for reducing the percentage of error, not an absolute control. If it is fed with cores more closely matched, the resulting swaged core will be more accurate. This is because the variation in pressure will be less with matched cores, and the bleed-off lead will be more nearly the same.

The cast core is better than the cut core because the process is more precise in control of weight. The extruded lead wire in theory should be more dense and less likely to have trapped gas, but this is negated by the inaccuracy of core-cutting. For the best of both methods, a cast core should be swaged in a core swage die, which gives both good original core weight control and pressure to re-form the core. However, using too small a core size for the die will only result in trapped air from folding of the core. The core should be reasonably close to the bore size. A .224 or a 6mm bullet's core will usually run from 0.186 for the .224 to 0.201 for the 6mm depending on the jacket, so a 3/16-inch core is not too bad a compromise. It is too large for making a 0.186 swaged core, being 0.1875 already, but only the most thick jackets such as early Gardiner .220 Swift jackets need small core diameters.

There is some disagreement about whether a tight-fitting core (swaged to jacket shape, which tapers internally), or a core with some air gap around it when inserted into the jacket is preferable. The first theory, which was advanced by B&A, claims that unless the jacket is an exact fit to the core before swaging, there is a potential for folding which can trap air. On the other hand, there is strong evidence that a tight-fitting core will trap air in the very bottom of the jacket, since it is simply pushed to the bottom and cannot escape as pressure is applied. Since there is no compelling proof either way, and the straight-sided core is much easier to make (taper dies require a fixed weight and punch diameter) and permits wide core weight ranges to be made, the somewhat more loose-fitting core is generally preferred.

One way to help ensure the expulsion of air from the bottom of the jacket is to put a dome in the core. One punch, usually the one in the die top, is ground with a precise cup shape which then swages a dome on the end of the core. The center of this dome touches the center of the jacket before any other part. As swage pressure is applied, the core flows from the center of the jacket toward the walls, rather than contacting all at once or the walls first. This irons the air outward and up toward the nose. The dome-shaped end of a core goes toward the bottom of the jacket each time for more uniform core seating.

Chapter 11

Core Length And Weight

Bullet cores can be cast in a mold, or cut from a spool of wire with a core cutter. Both methods are actually measurements of either volume or length, and not of weight. A core mold can cast either an adjustable weight, as with the ejecting type molds, or a fixed length which is then cut to the correct weight. The core cutter using lead wire has a stop that measures out a certain length of wire to be cut.

Since lead has a weight of about 2,868 grains per cubic inch, and a cubic inch is equal to the height or length of a cylinder times its base area, the length being enough to increase the area to be numerically equal to one, it is not difficult to formulate a chart showing lengths versus weights for a given diameter of lead wire. The area of a circle is 0.7854 times the square of the diameter. The circular section of a lead wire has an area equal to $0.7854(d)^2$, where d is the diameter of the wire.

If one were to use 3/16-inch lead wire for the example, this being a common size for .224 and 6mm bullets, it would be found that the cross-sectional area of the wire is 0.0276 in². In order to make a cubic inch, this wire would have to be approximately 36.23 inches long. (The product of the area and length is then equal to one.) It is known from the density of lead statistics that a cubic inch weighs about 2,868 grains, and thus a length of 3/16-inch diameter lead wire that is 36.23 inches long also weighs 2,868 grains. By constructing a chart of the lengths and weights for each size of lead wire, the handloader can accurately set the core mold stops or the lead wire cutter stops using a dial indicator or calipers.

The standard diameters of lead wire are as follows: 1/8-inch, 3/16-inch, 1/4-inch, 5/16-inch, and 3/8-inch diameter. The calibers generally used for each are dependent upon the jacket thickness. Usually 1/8-inch wire is used for the .17 and .20 caliber, while the 3/16-inch is used for both .224 and 6mm. A 1/4-inch diameter core is used for 7mm, .308, and 8mm in most cases. The calibers between 6mm and 7mm (such as .270 and 6.5mm) may use 3/16-inch wire but can possibly accept 1/4-inch in some thin jackets. The 5/16-inch wire is used for all the 9mm and .38 calibers, such as .380 ACP, .357 Magnum, and .38 S&W. The 3/8-inch wire is used for .41, .44, and .45 calibers.

On special order, in-between sizes of lead wire can be obtained. These include 5/32-inch, which is ideal for .224 jackets drawn down from 6mm jackets to make long, heavy-walled bullets. A 7/32-inch diameter is used for the .257 and .270 calibers, and sometimes for 7mm with thick-walled jackets. The 9/32-inch diameter might be used for thick-walled jackets in .358 caliber rifles or thin-walled 8mm caliber jackets.

A chart converting these fractional standard sizes into decimal sizes is given:

1/8	0.125
5/32	0.15625
3/16	0.1875
7/32	0.21875
1/4	0.250
9/32	0.28125
5/16	0.3125
3/8	0.375

The area of the cross section of each of these sizes is given in order:

0.01227 in²	1/8
0.01917 in²	5/32
0.02761 in²	3/16
0.03758 in²	7/32
0.04909 in²	1/4
0.06213 in²	9/32
0.07670 in²	5/16
0.11045 in²	3/8

Dividing each of the areas into one cubic inch gives the length to equal a cubic inch of lead of the specific diameter. This is the same as taking the reciprocal of the area (dividing into one). For example, 3/8-inch wire that is 9.0539 inches long will weigh approximately 2,868 grains, since it has a volume of a cubic inch. To find out how many grains an inch of 3/8-inch diameter wire would weigh, and thus get a constant of grains per inch for calculating length and weight, the density can be divided by the length required for a cubic inch. A chart giving the grains per inch for each diameter of wire has been calculated:

Wire Diameter	*Length per Cubic Inch*	*Grains per Inch*
1/8	81.4966″	35.190
5/32	52.1648″	54.980
3/16	36.2188″	79.185
7/32	26.6099″	107.779
1/4	20.3707″	140.790
9/32	16.0953″	178.189
5/16	13.0378″	219.976
3/8	9.0539″	316.770

Since the chart gives the grains per inch of each diameter of wire, moving the decimal place on the grains/inch constant to the left three places gives grains per thousandths of an inch. In order to use the chart with a density different from that of pure lead, simply divide the density of the lead given in pounds per cubic inch by the length per cubic inch. The result will be the grains per inch. Moving the decimal three places to the left gives the grains per thousandth of an inch, and this can be used to set the depth of a core cutter stop or core mold plunger using an indicator or calipers.

As an example, suppose that a bullet of .44 caliber weighing 200 grains was desired. First, the jacket weight would be subtracted from 200 grains. If the jacket weighed 30 grains, this would leave 170 grains for the core. Since the jacket would take a core of 3/8-inch diameter, this size would be selected in the table and the grains per inch used to determine how long a core to make. If one inch gives a weight of 316.770 grains, then how many inches would give 170 grains? The answer is found by dividing the desired weight by the weight per inch. The answer is 0.537 inch. This makes sense since we know that 170 grains is about half of 316 grains, and of course it would take about half as long a piece of lead to weigh half as much.

Setting the depth of the cavity in the mold, or the length of the space between the stop and the shear line in a core cutter, will result in making a core that weighs very closely the same as the calculated amount. The chart also helps avoid problems in selecting bullet weights for a given length of jacket or style of bullet, since it is known how much an inch of the core will weigh just from looking at the chart. It would be impossible to make a bullet shorter than about an inch in .375 caliber, as an example, that would weigh over 300 grains, since a plain lead .375-inch diameter slug weighs 316.77 grains per inch. And it is obvious that a .224 caliber bullet in open-tip style could not be made to weigh 100 grains unless it uses a jacket at least an inch long, since a 0.21875-inch diameter core alone will weigh 107.779 grains per inch.

Alloys of lead will weigh slightly less per inch, although the figures for area, length per cubic inch, and conversion to decimal numbers are not affected by the alloy selected. These figures are given in full so that the handloader can make use of them in other calculations, such as the pressure in the die figured from press gain and punch tip area, or to quickly see if a certain size core will fit into the jacket selected.

Chapter 12

Bullet Jackets

Until the development of smokeless powder, in 1885, a bullet made of lead was suitable for all calibers of guns. With smokeless powder and the resulting higher velocities and smaller bores, the heat and torque against the soft, easily melted lead bullets began to result in lead fouling the barrels and tearing of the bullet as it slipped against the rifling. Various attempts to insulate the lead from the barrel and the hot gas included paper patching, which was successful even for some military calibers using semismokeless powders, as long as the velocity was in the range of 2,000 fps or less.

Lead was used for bullets simply because it was easy to cast or swage, cheap, and had sufficient density to permit good ballistic performance. The first smokeless military rifle caliber in use by a major power was the 8 × 51 R French Lebel of 1886, which was called the "Balle D" load in our country but the "Balle M" in France. This cartridge was loaded with a solid bronze bullet containing no lead! It was successful, but was eventually replaced with lead core bullets for longer range. The bronze Lebel bullet was also the first military spitzer, all other major powers still using the round nose shape at the time. These improvements helped to make up for the relatively inefficient bullet material, but in general it was evident that only lead had all the necessary qualities for obtaining maximum performance in small arms bullets.

Actually, there was a modern style of jacketed bullet developed about 1874 by Lieutenant Colonel Bode of the Prussian Army. He devised a bullet with a copper jacket filled with lead. The copper insulated the lead from the bore and hot powder gas, gave a tougher substance for the rifling to grip, and did not tear off or smear to foul the bore as badly as lead. Copper jackets were further developed in the following years at the Swiss Federal Munitions Factory in Thun. Among the experimenters were Hebler, Lorenz, Schmidt, and Major Rubin. Professor Schmidt and Major Rubin combined efforts to produce the 1889 7.5 × 53 MM Schmidt-Rubin rifle, which used a paper patch bullet and semismokeless powder. Prior to the development of the French Lebel in 1886 were several copper-jacketed bullet loads and cartridges, notably the 8 × 53 R Schmidt-Rubin and the 8 × 53 Hebler-Lorenz, both produced in 1885 but not adapted for standard military use. Both used black powder.

Copper makes a good jacket for specific performance requirements, such as good expansion without shattering in a large game bullet. Being relatively soft, it does tend to strip across the rifling with high acceleration loadings, and it leaves a thin wash in the bore which can react with moisture to produce corrosion by electrolysis action. Solid copper bullets have also been used with success in very fast subcaliber rifles, such as the .17 caliber. Generally, better performance is obtained with jacketed bullets using lead cores in these calibers as well as larger bores. Copper wire, with the tip drilled and filled with lead, has been used in very small bore rifles. The copper wire can be swaged in regular home swaging dies in these small sizes because of the high pressure it is possible to develop in the small die bore.

Copper washes or platings on lead bullets are used even today on rimfire bullets, but in general the accuracy of a lubricated lead bullet is better and there is little difference in the fouling produced. At higher speeds, where there might be an improvement by the use of copper as far as fouling reduction is concerned, there is not enough strength in the copper wash to resist stripping across the rifling. Another use of the thin wash was over the old copper jackets used on sporting ammunition in the early days. This wash was of tin, which was very thin and could be rubbed off with a fingernail. It was apparently there to keep the ammunition looking factory-new, since the copper or gilding metal bullets might take on a tarnish after sitting for some time on a dealer's shelf.

The U.S. government experimented with tin-plated bullets at the Frankfort Arsenal, in an effort to reduce the fouling that was caused by cupro-nickel jackets. Cupro-nickel was used as a jacket material in the United States and England in order to produce a jacket tougher than copper. It consists of from 60 percent to as much as 87 percent copper, with from 40 percent to 13 percent nickel. The color is silver-gray in spite of the large amount of copper. Although admirable in most ways as a jacket material, cupro-nickel tended to leave actual lumps or smears near the muzzle, where velocities reached their highest point in the bore. These lumps were so difficult to remove that a corrosive ammonia solution was poured into the bore, which was plugged and allowed to sit while the ammonia dissolved the fouling. Unfortunately, the cleaning solution also attacked the bore under certain conditions. In 1920, the tin-plated bullets were manufactured by electrochemical plating over the cupro-nickel. This did solve the fouling problem, but the tin began to pressure-solder itself to the brass cartridge necks. Several rifles were wrecked and the practice was discontinued.

In Europe, a very mild steel has been used as a jacket material with good results. It does tend toward shorter bore life, but to improve wear the steel is plated with copper or cupro-nickel. England used a nickel plating, which gave a spotty type of fouling. The Western Cartridge Company developed a gilding metal alloy called "Lubaloy," which was copied by Nobel and called "Nobeloy." This was one of the more effective kinds of platings over other jacket materials, and was eventually matched by other bullet makers in improved alloys of gilding metal. Today, the best and most common jacket material in use is a gilding metal formula. The basic mixture is from 90 to 95 percent copper with from 5 to 10 percent zinc. Lubaloy is gilding metal with a small amount of tin added.

Even gilding metal does tend to foul the bore slightly, but the amount and type of fouling are completely different from earlier jacket materials and are suitable for nearly any speed and caliber of bullet. A thin wash is applied down the bore, like a plating. The amount of thickness is almost microscopic in most cases, and does not tend to pile up in lumps like cupro-nickel. It is less active in the presence of moisture than plain copper, as far as corrosive action on the barrel steel is concerned, and in general is completely satisfactory as a jacket material.

Bullet jackets normally start as a long strip of gilding metal in a huge disclike roll. The rolls are fed through automatic feed machinery into punch presses, where discs are blanked out, and pushed through drawing dies to form cups. The cups are drawn in several stages until they reach the proper length for trimming. Usually, the edges of the jacket are very uneven and one side may be thicker than the other at this stage. It requires the greatest diemaking skill, constant attention, and quite a bit of luck to obtain jackets with perfect wall concentricity from any given run of jackets. The finest benchrest quality jackets, having extremely good concentricity and thus jackets drawn very evenly in wall thickness, may cost twice as much as a jacket run on the same machinery, of the same materials, for use in hunting rifles. The difference is in the amount of setup, inspection, resetting, and other constant bother to maintain high-precision tolerances.

After drawing, the jackets are trimmed to length. An untrimmed jacket usually has a considerable amount of feather-edged roughness and is uneven at the end. Trimming can be done by means of a lathelike machine, with a tool bit that presses against the jacket as it is held in a collet and spun, or in a pinch-trim die that actually forces the jacket through a hole in which a plug or punch takes up all the room after fixed distance. The metal is pinched off by the action of the plug slipping snugly into the die. A rotary-trimmed jacket may have a slight roll inward at the jacket mouth, which must be swaged out using a tapered plug punch, or it may be straightened by the manufacturer. It will usually be thicker at the end since it is cut off at a right angle to the wall. The pinch-trimmed jacket is normally cut off with a long angle, being sheared by a forcing action, and thus has a thinner edge. The main difference to the home swager is that some dies will not eject pinch-trimmed jackets as easily, especially if they are rough or require extra force to free the jacket from the die walls. But pinch-trimmed jackets can make very precise bullets, usually having better control over the jacket length tolerance than the rotary-trimmed type. There is little to be concerned about in the matter of how the jackets are trimmed, as long as they are suitable for the dies and the purpose.

In recent times, the bullet manufacturing firms have stopped supplying jackets to the home swaging trade, since the bullet manufacturing machinery has gone to fully automatic, multistation presses that take in gilding metal and lead, and turn out finished bullets. In the earlier times, when separate presses were used for making jackets and swaging bullets, runs of jackets were sometimes set aside for sale to handloaders. Today's market source for jackets is usually the custom swaging component supplier or custom drawing firm, which specializes in jackets for resale rather than its own bullet production. Jackets are very similar to other common drawn products, so that as long as there is a demand, the manufacturers of flashlight cases, blasting caps, electronic component cases, and similar deep-drawn

A rotary bullet jacket trimmer with automatic feed.

Close-up of a jacket being trimmed in the rotary trimmer.

metal containers can be persuaded to supply a minimum of a million or so jackets on a special order at a reasonable price.

Jackets can also be drawn at home, but the effort is so great that it does not pay to start from flat stock. A number of draws are required, and the skill in operation is almost as great as the skill in making such dies. Existing cupped jackets and components such as .22 rimfire cases, shotshell primers, and other drawn items can be redrawn easily, in one or two steps, to make certain calibers of jackets at home, however. Once the metal has been cupped and formed, it is not difficult to adjust the diameter or increase the length slightly. Copper tubing can also be used at home without great difficulty, as long as the bottom is not closed completely. This operation requires considerable pressure and stroke, more than the reloading press can usually supply for most calibers. Other methods, such as the use of a smaller, internal gas check or a pistol half-jacket to close the base, make a protected-base, copper-tube bullet practical.

It has been suggested by many home swagers and experimenters that tools be supplied to draw jackets out of aluminum can material. At first, this appears reasonable, since aluminum draws very easily and the metal is available in sufficient quantities almost for the asking, in the form of nonreturnable aluminum cans. However, it would soon be discovered that aluminum melts at a much lower temperature than any of the common jacketed materials, and fouls the bores more readily. It is also rather soft and does not resist the torque of the rifling, and upon being worked becomes brittle and may shatter instead of expand. In short, it is not suitable. Steel jackets are more reasonable, but making jackets out of "tin" cans is a hard and unrewarding job requiring a considerable investment in dies and labor. With copper tubing and commercial jackets so low in cost and readily available in most diameters of interest, and with redrawing dies so simple to use for producing calibers one or two sizes smaller than standard commercial diameters, it doesn't make sense to attempt to draw jackets from flat stock at home.

One of the more interesting jacket materials is a sandwich stock made of mild steel with cupro-nickel or gilding metal sheets, like a thick foil, pressure-bonded to both sides. This has been used in Germany, and a commercial version is used in some Norma bullets. The material appears to be very successful, without increasing bore wear any noticeable amount. It is, of course, a little more expensive than plain gilding metal and is not used in the United States for jackets for resale. A large quantity of short .30 carbine jackets were produced and sold as surplus from the days of the M1 carbine, which used steel jacket material plated with either copper or a gilding metal alloy. These look like gilding metal jackets but are attracted by a magnet. The only problem with such jackets is that during point forming in the swage die, they do not point up as sharply without little folds or wrinkles appearing in the ogive section. A little more pressure appears to be necessary to form them. However, they do not seem to cause any problem to the swage dies or to rifle bores.

In the early days, bullet jackets were often made only two- or three-thousandths of an inch thick and were almost the same thickness from top to bottom. Today's jackets taper from mouth to base, getting as much as 0.040 inch thick at the base in some calibers and tapering to as little as 0.005 inch at the mouth. Very thin jackets are seldom found in rifle calibers today. Until recently, it was difficult to find a handgun jacket in the smaller bores, such as .32 or .30, that would expand at normal velocities for such handguns. Tapered-wall handgun jackets with thin walls and thicker bases have been produced in .38, .44, and .45 calibers, and these can be redrawn to smaller sizes with good results. In the .25 ACP caliber, it is possible to use fired shotgun primer cups for jackets, after a two-step drawing operation. And when a very explosive bullet jacket is desired for varmint shooting or low-velocity loads that will still expand, the .224 caliber, .17 caliber, .20 caliber, and 6mm caliber are all served well by using fired rimfire cases properly drawn and formed for jacket use.

When a jacket is reduced in diameter, it becomes either thicker in the walls or longer, or perhaps both. Thus, if a handloader wishes to produce a bullet with thick walls for use at high velocities, he might choose a jacket that is one caliber larger than the rifle bore and then have a die made to reduce the jacket one caliber. An example would be a .220 Swift being used at moderate ranges where ordinary jackets might be too fragile for the velocity on impact. A 6mm jacket might be drawn down to .224 caliber in one pass through a ring die. The resulting jacket would be ideal for the higher speed since it would have much thicker walls than normal. Provided only one caliber size reduction is taken, and provided that the jacket is not lengthened any great amount at the same time (a function of the clearance between punch and die wall), the pressure required to do this is very low. An ordinary loading press will handle the job. The quality of the jacket is not impaired in the operation, again provided that one does not attempt too much reduction in diameter or increase in length in one pass.

Using the same example, a longer than normal jacket can be produced for fast twist heavy bullets, like the .224 Clark. The difference is that the punch and die clearance in the ring die or jacket drawing set will be smaller, so that the walls are stretched instead of thickened. Pressures are greater, diameter change can be the same as in the prior example, and quality of resulting jacket can be the same but may suffer from wall thickness variation if too great a lengthening is attempted. Since the next larger calibers usually require slightly longer jackets anyway, this is not a great problem. Some jackets do not draw as easily as others, so a handloader would be wise to specify the jacket and to make certain of a supply of the proper one to use with the dies when they are produced.

Chapter 13

Rimfire Case Jackets

In the early days of centerfire small caliber rifles, there were no jacketed bullets available to experimenters like Harvey Donaldson, and it was necessary to produce a jacketed bullet at home for the .22 centerfire. The copper cases used for early .22 rimfire cartridges were nearly the ideal length and diameter for jackets. During the 1920s Donaldson and others devised simple hand- or press-operated drawing dies and punches to iron the rim out of the fired cases and thus produce a bullet jacket.

After centerfire smallbores began to become more popular, bullets were offered by commercial manufacturers and the making of bullet jackets at home lost popularity. However, during the World Wars, when bullet jackets and in fact all reloading components were difficult to obtain, there were rebirths of the rimfire case bullet jacket for .224 caliber rifles. (By .224 caliber is meant all the various designations of .22 centerfires. This would include the .222, the .223, etc. Today, nearly all .22 centerfire rifles use a .2240 diameter bullet regardless of the actual name of the cartridge.)

Bullets were so produced by Fred Huntington of the Rock Chuck

Bullet Swage firm, by Vernon Speer of Speer bullet fame, and by countless small operations which produced either the tools to make such bullets and jackets at home or made the bullets for resale. Usually, whenever a supply of commercial jackets became available, or when gilding metal rolls were available to bullet manufacturers, the rimfire jackets were dropped. Because early .22 rimfire cartridges used a soft copper case, they were easily swaged into jackets. But there were two problems with these jackets that gave the process a bad reputation.

First, the rimfires are not reloadable, and so there was no reason to worry about the use of mercuric priming compound. Mercuric priming is very injurious to copper and brass, making them brittle after the primer has been fired. Centerfire reloadable cases were loaded with nonmercuric primers when reloaders complained of the high rate of case failure, but of course there has never been a practical way to reload rimfire cases at home. Since rimfires were not reloaded, and the mercuric priming was very positive and powerful, such priming was used even as late as the 1940s. Rimfire cases that were made into bullet jackets often were so brittle from mercuric fouling that they would shatter or fly apart in flight, causing poor accuracy and unexplained fliers. Soft copper jackets used at high speeds would strip in the rifling, and deposit copper fouling in the bores. Neither of these faults aided accuracy of rimfire jacket bullets. Today, however, the situation has changed.

Mercuric priming is no longer used. The cases are not made of soft copper, but an alloy that is more nearly like gilding metal. The brass contains a higher percentage of zinc than gilding metal, and thus is less likely to foul the bore but does not draw or expand as easily on impact. Since the walls of jackets made from rimfire cases are almost half the thickness of commercial bullet jackets, it follows that they will engrave more easily with less friction, and thus if anything are more easy on the bore than commercial jackets. There is no reason to worry about the use of rimfire cases for bullet jackets. They will not harm the bore of a rifle.

The main reasons commercial jackets are not made of the 30 percent zinc alloy is that the drawing of the jacket would be more difficult, and the jacket would not perform nearly as well in mushrooming. Rimfire jacket bullets in .224 or 6mm (the only practical calibers which use rimfire cases as jackets, other than the .20 or .17 caliber, which can use drawn .22 short cases) are very explosive and should not be used for any medium or large game. They are strictly target and varmint bullets. On small varmints they do a remarkable job. Usually, they will expand completely to the base even at very moderate speeds. This is just as well, since the thin jacket will not take extremely high rotational speeds that accompany the 3,500 fps and higher varmint bullet velocities.

A rimfire bullet is an experience to the varmint hunter who can take his shots in the ranges below 300 yards and use a little slower velocity loads. The effect of this thin jacket on varmints at 2,500 fps is very much like a high-speed bullet from a .220 Swift. It is difficult to believe that a .223 or .222 centerfire caliber using not over 3,000 fps muzzle velocity can produce such explosive results at moderate range. Of course, the main advantage would be lost if the bullets were not accurate. Tests of rimfire bullets by many experienced benchrest shooters have given results from poor to excellent, comparable to regular commercial bullets. The poor results usually came about when an inexperienced bullet swager attempted to go for accuracy right away before learning the fundamentals of swaging. However, some rifles, especially in 6mm caliber, do not shoot well with the lower-velocity thin-walled jacketed bullets, and some tinkering with the load and bullet weight is necessary. Whenever the velocity is held down to a reasonable level and the bullets are made according to the same good technique as might be used with benchrest-quality jackets, the results have been comparable to factory-made bullets.

One of the most attractive things about using rimfire cases for bullet jackets is the lack of cost for materials. Rimfire jacket material is simply picked up off the ground at most shooting places. Material that would otherwise go to waste can be formed into bullets that would cost five or six dollars a box. The operation takes but a single die—the operation of drawing the rimfire case into a jacket, that is. The same dies that are used with any commercial jacket are also usable with rimfire jackets, provided that a slightly larger core-seating punch is utilized. (The thinner the jacket, the larger the diameter of the core-seating punch must be. This is true with any jacket.)

A 3/16-inch lead wire core will make a good core for a rimfire jacket, although for more precise results it should be swaged to about 0.201-inch diameter. This is larger than the size recommended for commercial jackets; so a commercial jacket core swage would actually be a little small for use with rimfire jackets. It happens that most 6mm commercial jackets require a 0.201-inch core. Thus, a shooter who has both 6mm and .224 rifles can use his 6mm core swage for .224 rimfire jacket bullets. The usual commercial jacket core size for .224 caliber bullets is about 0.186 inch in diameter.

The process for forming a rimfire jacket is easy. First, the jackets

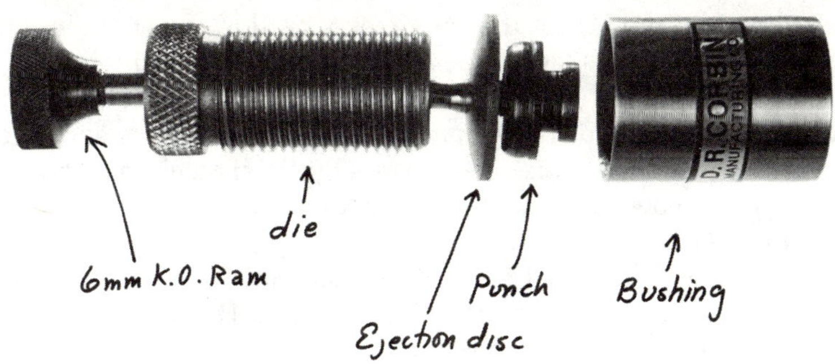

A rimfire jacket-maker die and punch set.

should be sorted into brands all of one kind. Then, all the same brand of jackets should be kept together and washed thoroughly in hot, soapy water. After complete drying, they are individually picked up using the fingers which have been moistened with a drawing lubricant, such as lanolin and castor oil or Draw Die Lube, and placed over the punch of the jacket-drawing die. The die is adjusted in the press so that the case rim will enter the mouth of the die and draw off into a smooth cylinder just at the top of the press stroke. This requires very little effort. If the die is set at random, it will probably take far too much muscle power to run the case far enough into the die to draw off the rim. About half or even just a third of the case should be run into the die before the ram goes as high as it can.

The jacket can be ejected from the die and set aside for later, to be run through with the die set a little lower, or the exact point can be found where the mouth of the case will just enter the die and be supported enough that another case can be pressed in behind it. The drawing die is open at the top, and after three or four cases have been ironed and pushed into it, the first one will pop out the top. This adjustment takes a little fiddling, since it is more than half of the case length into the die and will start getting critical on pressure required. If the die is set too high in the press head, far too much of the case will be left unsupported below the die. Another case placed over the punch will just bend the unsupported mouth of the case in the die.

When the jacket tends to stick on the punch, instead of coming off and staying in the ring die, an ejection system made of a disc slipped over the punch and a bushing that rests on the bottom of the press throat and goes around the ram will push the jacket off the punch on the ram's downstroke. The assembly is: bushing slips over the ram, punch slips into the slot of the ram, disc slips down over the punch. Then the jackets are placed one at a time over the punch, mouth down, and pressed up into the ring die. The ring die is a drawing die with a proper shape and size of opening that closes to a very accurately lapped hole. The case must be drawn slightly smaller than original diameter to remove the rim fold smoothly and also to fit into the core-seating die.

It has been found that a minimum die constriction diameter of 0.218 inch will result in about a 0.219-inch O.D. jacket. A mouth opening of about 5/16 inch allows the case rim to align with the bore of the die. A very smooth radius with a fine finished surface will permit the most easy drawing of the case into a jacket. The rim is not folded but is actually straightened out and drawn smoothly into the walls of the jacket. The firing pin dent is still visible but is very nearly gone after the lead core is seated. The location of this dent hardly affects the grouping of the bullets, and its effect on accuracy can be disregarded for normal shooting activities provided that it is not a tear completely through the jacket wall. Very sharp firing pins can cut the case rim badly enough to spoil their utility as a bullet jacket.

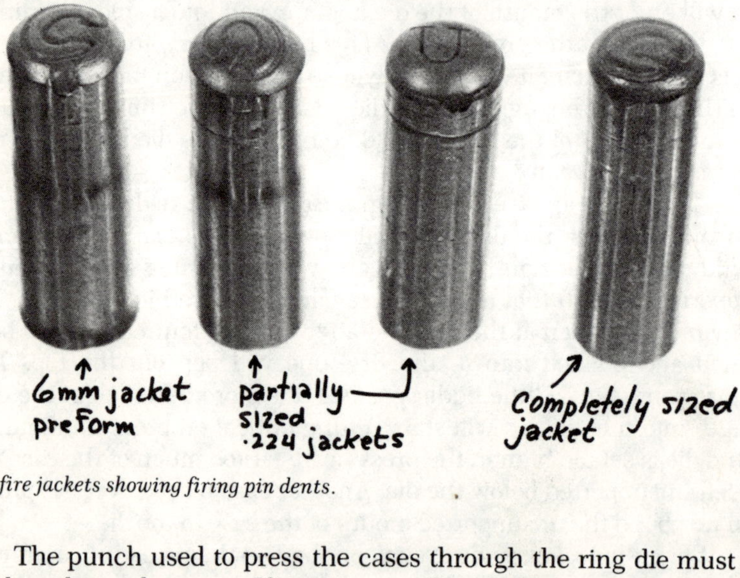

Rimfire jackets showing firing pin dents.

The punch used to press the cases through the ring die must be held to close tolerances. If it is too small, or becomes worn from the abrasive action of the firing residue, there will be a crease in the head of the drawn case where the rim should be ironed flat. Eventually this will become a fold, and the accuracy of bullets made with these jackets will suffer. The proper diameter is 0.199 plus or minus one-thousandth of an inch. This is for use with .22 short, long, or long rifle cases. The .22 WMR and .22 WRF cases require a different die and punch. They are too long for any but the most heavy 6mm bullets in any case. Even so, they must be expanded slightly to fill the bore of the core-seating die.

In using .22 long rifle cases for 6mm jackets, the mouth is expanded so that it will fill and seal the bore of the core-seating die. Only lead-tip bullets can be made for two reasons.

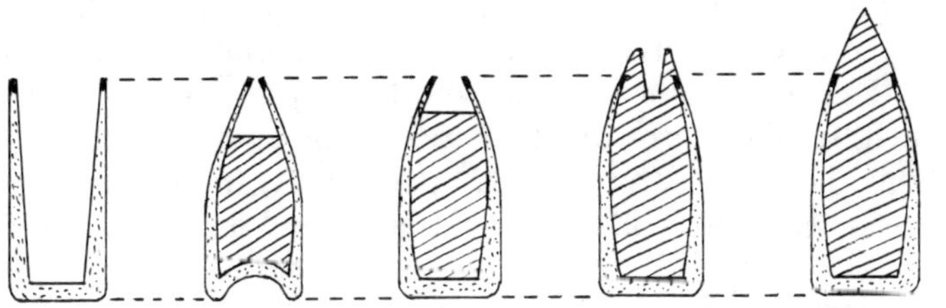

ILLUSTRATION of how the bullet shape, base, tip style, and amount of core seated within the jacket or exposed above the end of the jacket actually determines the bullet weights for that jacket length. The short ogive (rounder nose) makes the heavier weight bullet in the same styles. Spires make the lightest bullets, spitzers next, round noses the heaviest if all other features are kept the same.

Bullet shape and core determine bullet weight.

First, the diameter of a fired .22 long rifle case may be about 0.225 depending on the chamber and kind of gun in which it was fired. This is too large for a jacket for .224 caliber, and is drawn down to about 0.219 to remove the rim and creases. It is already far too small for a 6mm bullet jacket, unfortunately, even before the rim is drawn away. However, if it is also reduced to 0.219 inch in diameter and then the mouth is given a slight bell shape to make it approximately .243 inch just at the edge, it would be possible to use a long lead core and apply pressure to this core with a punch as large as the core-seating die bore.

A seal would result between the punch on one end and the edge of the jacket at the other end. The lead core would expand until it began moving the body of the jacket toward the die walls.

Eventually, the jacket would become the same diameter and shape as the core-seating die walls. This could only happen if enough lead was left protruding so that the full-size core-seating punch never actually came down against the jacket. If it did this, it would apply pressure to the jacket walls directly and fold them. It so happens that a .22 long rifle case is a bit too short anyway for a 6mm open-tip bullet; so a practical rifle bullet using the 0.700-inch length rimfire case (after drawing) would need a lead tip anyway for weight. About a 65-grain bullet is formed using a significant lead tip exposure in a .22 long rifle case, in 6mm caliber.

In the .224 caliber, an open-tip bullet of about 52 grains is a very good match to the long rifle jacket. As light as 35-grain bullets can be made in open-tip style in a .22 short case, or 45 grains with lead tip. A 65-grain .224 bullet can be made with lead tip in the long rifle case. The lead tip bullets are much more explosive, even with standard commercial jackets, than the open-tip styles. As the rimfire jacket is very thin anyway, the combination of lead tip and rimfire jacket means velocity should be held down far below standard levels and yet the effect on impact will be just as great as with hot loads and standard bullets. Thus rimfire jackets not only save on bullet component costs, but force the shooter to save wear and tear on his bore, use less powder, and work toward what is usually a more accurate range of velocities in his rifle. Maximum speed enthusiasts will find little of interest in the rimfire jackets, except for the effect of high-speed impacts.

Much has been said about annealing the jackets made of rimfire cases. With modern jacket materials, this is only necessary if the ogive is brought down to such a small tip that the jacket material begins to fold over on itself instead of drawing in smoothly. There is no other particular reason for annealing. The amount of heat and time necessary is just found when the jacket will make a bullet without folding up at the ogive. Usually, heating the jackets to 500° as a stress relief rather than an actual anneal just until they are brought evenly to that temperature without cold spots, a matter of minutes, is sufficient. Since the various jackets use different materials and have had different work-hardening histories, there is no specific time and temperature that will give uniform results on all jackets. Heat a few minutes, and make some bullets. If they form up well, the job is done. If they still buckle, heat a little longer or a little hotter. Avoid heating to the point where the jackets become discolored and begin to take on dents and wrinkles in the midsection. This is too hot and means the jacket material has been ruined. At 800°F the destruction begins with most alloys tested.

Prior to drawing, there is no need to anneal. A good washing is all that is required. After drawing, the lubricant should be completely washed off. Draw Die Lube is soluble in water. In hot water, it makes its own cleaning agent. The jackets should be washed in hot water and then spread out so they will dry quickly. Any lubricant or water that is in the jacket will stay there when the core is seated and unbalance the bullet. Lube on the outside of the jacket is a necessity, and is applied fresh by handling the jackets just prior to core-seating. Lube on the inside is a mistake.

It has been suggested that larger-caliber bullets can be made using rimfire cases. This is true but not practical for two reasons. First, the length of jacket is too short for practical weight bullets over 6mm in diameter. Second, the amount of swaging upward is excessive when calibers over 6mm are attempted. The result is a bullet that has rings and uneven mouth, a small base, reverse taper so that the middle portion is larger than the base, or other problems that come from attempting to go too large with a jacket in one step. If carried too far, the mouth will split from excessive belling. The .22 WMR (rimfire magnum) can be used for up to .25 caliber bullet jackets. These cases are not as easily found and the process is not usually recommended for this reason. It is possible to make .25 ACP bullet jackets out of fired .22 cases, but it is even easier to make these bullets using spent shotshell primer cups. Nearly any trap club has a handloader who is glad to donate plenty of spent shotshell primers. The tooling is available to turn these into very satisfactory .25 ACP jackets.

Chapter 14

Matching Dies and Components to Jackets

The jacket is possibly the item most overlooked by a beginning swager, and generally the last item purchased after the dies, core molds, and other accessories of swaging. Since it is the jacket that determines the diameter of the lead core, and the over-all length of the bullet, as well as the diameter of the core-seating punch, it would be a good idea to purchase jackets with the swage dies and not afterward, except as replacements. To select a jacket to use with a given set of dies is not difficult, but it is better to have the dies made to work with a selected jacket in the first place.

Starting with a bullet weight and caliber, one must decide what nose shape is desired. If a wide range of weights is desired, then one must decide upon a nose shape that will permit the lightest of the weights to still have about 50 percent of the bullet length in the shank. After deciding whether the bullet desired is physically possible, in regard to diameter, weight, and shape, a jacket should be selected that will come close to meeting the exact requirements of this bullet length. The length of jacket for a given weight will depend on caliber, shape, and to a larger extent, whether or not the bullet will have an exposed lead tip, or an open tip.

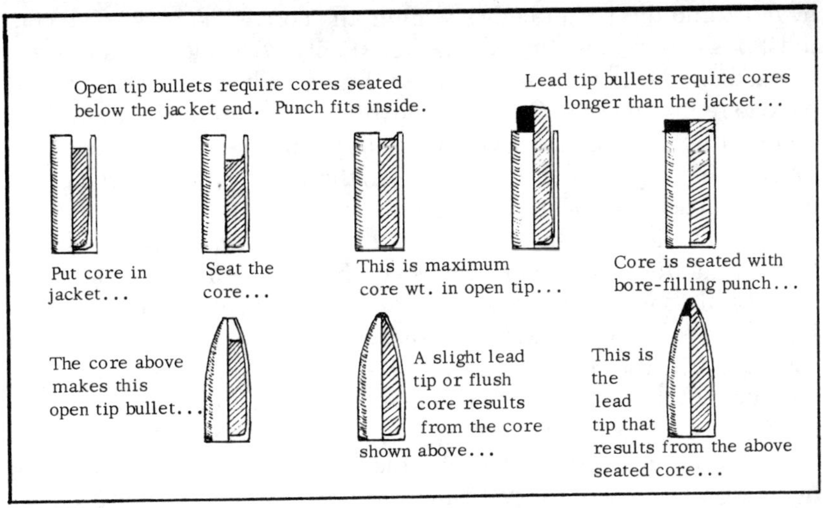

Core weight affects bullet style.

A jacket which makes a 150-grain open-tip bullet in .30 caliber might well make a 170-grain lead-tip bullet in the same shape using the same die set. A bullet jacket that makes a 168-grain open-tip bullet with a boattail will usually be far too long to make a suitable 150-grain bullet of any flat-base style. The ogive can be left partly empty without harming the accuracy; most fine benchrest bullets do have a slight vacant area behind the bullet tip, which is formed of jacket only. This is one way that a single jacket length can serve for several weights of bullets. But only about a third of the ogive can be left empty of lead before problems with balance and with proper forming of the jacket in the point-forming die begin to occur.

Generally, the diemaker, given the proper information, can suggest a jacket to suit the needs of the handloader. It is necessary for the diemaker to have a definite jacket specified in order to make the core swage die and the core-seating punches. Since jackets differ in their internal diameter and the rate of taper, a lead core that is just large enough to slip into a thin-walled jacket will not fit into a thicker-walled jacket. The core swage has to be large enough so that the lead wire or core from a core mold will just slip into its bore, but it cannot be so large that the resulting core will not slip into the jacket easily.

If a handloader purchases a set of dies for a given jacket, and later tries another jacket with different wall thickness, it is very likely that the cores will not be a proper fit. Any time that a core must be pressed by force into the jacket before seating, the core is too large. The core must be an easy slip fit into the jacket. If only a few such jackets are to be used, then lead wire can be pulled through a hole in a piece of steel plate to reduce its size. Generally, pushing cores through a draw die to reduce their diameter is not satisfactory, since they will bend and mushroom at the base rather than push through the die. Draw dies are most successful on jackets and on finished bullets.

The diameter of the core-seating die is most closely related to the final size of the bullet, not to the jacket. Jackets are always undersized by several thousandths from the final bullet diameter desired. If a jacket is already at the final bullet diameter, then there is no way it will slip easily into the core-seating die. Pressing a jacket into the die under pressure as it is squeezed down to fit only means that the jacket will try to grip the die walls rather than spring back away from them, when the core-seating pressure is relieved. This causes hard ejection, and also helps to make a loose core inside the jacket, which is not good for accuracy.

The core-seating die is usually about a half-thousandth of an inch, or a little less, smaller than the bore of the point-forming die. These two dies can be made and tested with a jacket different from the one which will be used by the handloader, except for the fact that the core-seating punch must fit very closely to the inside diameter of the jacket that will be used. Thus, if no jacket is specified or ordered, and the dies are made with a stock jacket used for testing, there is a good chance the core-seating punch will be too large or too small for the jacket that the handloader finally decides to use. This is no great problem since punches are not expensive and can be made just by having the jacket, not the entire die, at hand. But a loose-fitting punch will mean poor pressure and a badly seated core, with bleedby. A tight punch will scrape the jacket walls. The proper size punch just barely marks the walls as the core comes up to final position, or at worst it may have a faint edge of lead just beginning to bleed by.

A simple way to test for length of jacket is to cut a length of wire about the same diameter as the bullet to the same weight as the bullet desired. If a lead-tip bullet is desired, a jacket the same length or slightly shorter than this lead wire will be needed. When the ogive is formed, some of the lead will be moved forward. If the diameter of the wire is much less than the diameter of the finished bullet, allowance must be made for the shortening of the lead core when the core is seated and expanded to the inside diameter of the jacket. If an open-tip bullet is desired, the jacket must be slightly longer than the core or sample lead wire. An exact length is not critical unless a very precise

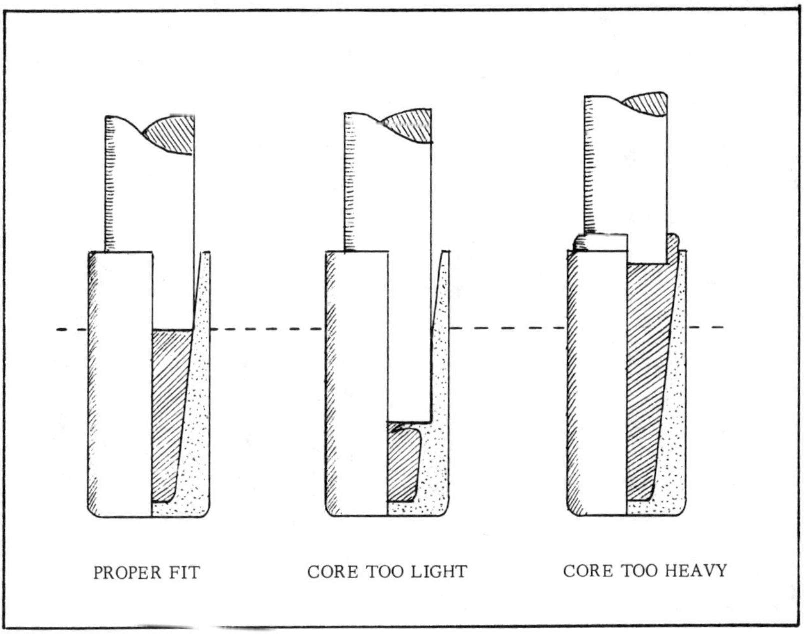

The core-seating punch with taper wall jackets.

weight is coupled to a very specific amount of exposed lead. Usually the weight is not nearly as important to any of the performance characteristics as the shape, care of manufacture, and style of bullet. Thus, the weight becomes the factor that is varied to gain the desired amount of exposed tip and to use the jackets which are readily available. Jackets are generally made in a few standard lengths in each caliber, not cut to order except for a rather high cost. This is not very important since the weight can be changed over quite a range without interfering with the goals a handloader might have in mind for performance. Or, alternatively, the weight can be fixed but the position of the core within the jacket varied without causing much difficulty in performance. These allowances mean that one or two jacket lengths serve most handloaders very well for quite a range of bullets.

In the larger bores, there are almost no jackets available commercially at a reasonable price or quantity. Any caliber is available if a shooter wants to purchase around a million jackets, since most drawing firms will set up to make this kind of quantity on special order. The less popular a caliber, the more likely it is that the swager will have to make his own jackets, either from existing caliber jackets or copper tubing. Whenever possible, it is cheapest and best to use

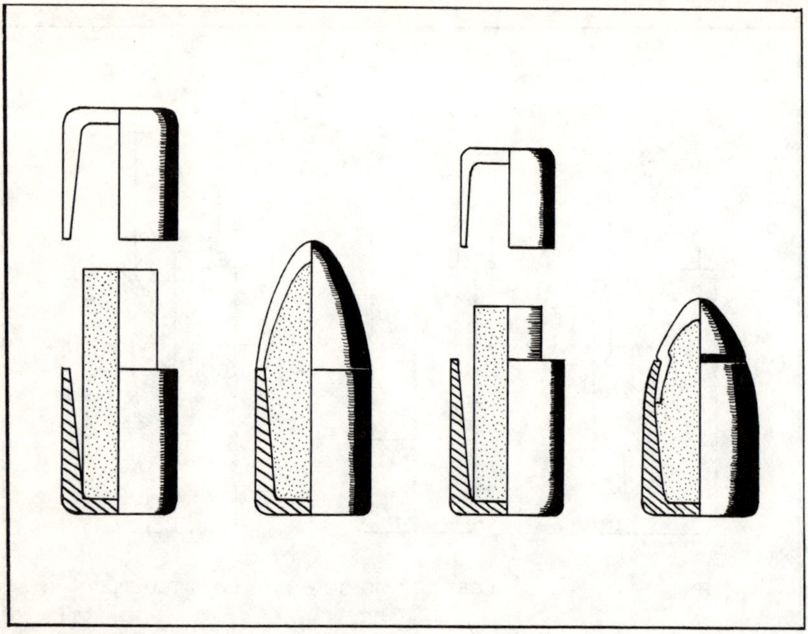

Multiple jacket bullets using two handgun jackets.

existing jackets. One way is of course to draw down a larger caliber. In the calibers over 8mm, there are few rifle-length jackets available to draw down. For lighter bullets, one can draw a handgun jacket up to about 5/8 inch in length, since there are half-inch-long .44 and .45 jackets available. Most of the standard weights for the calibers over 8mm require at least a one-inch jacket, however.

A simple way to make inch-long jackets in the larger bores is to place a pair of handgun jackets end to end over a lead core that just permits them to touch edges after swaging. The weight is fixed by the length of the jackets although by using combinations of half, three-quarter, and full-length pistol jackets several weights can be swaged. The resulting bullet looks very good, and has the unusual characteristic of being fully jacketed with no openings on either end, yet with the ability to expand in the center. This bullet should not be confused with full-jacketed military type ammo. The thin edges of the handgun jackets will mushroom rather well on impact, and the bullet performance can be quite good if somewhat unusual. This is a simple and cheap way to make .275, .358, .400, and .458 caliber rifle bullets and is well worth a try since nothing more than a quantity of handgun jackets is required to try it in a set of swaging dies. The lead length must be exactly correct so that the jacket edges just touch after the bullet is swaged. This is found by making four or five bullets until the proper length is found. If the core is too short, the jackets will fold. If too long, there will be a band of lead between the two.

The use of .30 carbine cases for .375 caliber bullet jackets has been tried and they make serviceable if not particularly accurate bullets. A pair of .38 handgun jackets end to end makes a much nicer and more precise bullet. The .357-inch jackets swage up nicely to as large as .375 caliber. Jackets can be expanded a considerable amount just by the pressure of the lead during core-seating. This is the way that a .22 rimfire case is made into a 6mm bullet jacket. In fact, the two techniques of either drawing down a jacket in a ring die, or swaging it up by flaring the mouth to seal the bore of the core-seating die and then seating a lead core longer than the jacket, give each standard size jacket at least three calibers in which it can be used.

Strictly speaking, a jacket cannot be "swaged down." The idea of swaging to reduce diameter is a misnomer; actually, reduction in diameter is done by drawing through an open-ended ring die, not by swaging. The process of swaging is a process of expanding upward from internal pressure, not reducing by squeezing through a constriction. Thus it would be difficult to use a larger jacket directly in a smaller swage die. The larger jacket would simply fold, or stick fast to the die walls. It must first be drawn through a die to reduce the diameter below that of the die cavity, then swaged back up slightly in the die.

Jackets have one other function in the process of swaging besides their benefits to the bullet itself. A jacket is more or less elastic, in that it will spring back toward its original size when pressure is removed. Lead by itself is an inelastic substance, which retains exactly the shape of the die into which it is pressed. An all-lead bullet will tend to remain in close contact with the die walls, whereas a proper size jacketed bullet will tend to spring back very slightly to original diameter when pressure is released. This lets the jacketed bullet free itself from the die walls. Generally, a core swage die will have harder ejection, requiring more force to free the lead core, than a core-seating die of the same size. This is because the core-seating die is used with the jacket and tends to free itself from the intimate contact between jacket and die as soon as the pressure is taken off the punches.

If the exact size of a die is to be measured, a soft lead slug is better than a finished bullet with a jacket because of the springback. However, the bore diameter of a die is not going to be the same as the bullet it will swage for the same reason. When making very precise diameters of bullets, it is necessary to use the jacket material, thickness, anneal, and length of the original jackets used to make and test the dies. This is another good reason to purchase jackets with the dies, and to specify when the dies are ordered exactly which jackets will be used the most with them. A set of dies cannot actually be guaranteed to produce a precise diameter of bullet when the materials of the bullet are not included in the specifications. Even differences in lubrication and technique will vary the diameter of the bullet by a small amount, so that the best guarantee is the fact that bullets can be made according to specification if everything is done correctly, with the same materials as the diemaker used.

Chapter 15

Positive Stop Versus Feel

Ideally, it should be possible to seat cores, swage cores, form ogives, or handle any other swaging function by setting the die at a fixed point and running the ram of the press the same distance forward each time. In practice, there can be enough variation in components to make this dangerous to the dies, since it is possible to develop pressures far in excess of anything a die can stand. Many shooters feel that it is best to gauge the pressure by a developed sense of pressure on the press handle. They call this seating or swaging by feel. Some of the top bullet swaging enthusiasts whose bullets win precision matches consistently are among this group.

Before going into the development of a feel for pressure, or discussing how to set up a positive stop, we need to make certain that the relationship of force, area, and pressure are well in mind. In bullet swaging, everything depends on pressure. There is no pressure developed within the die until a punch actually contacts the lead or jacket and meets resistance from it. If the resistance is slight, no more pressure can be developed than the resistance to it. Of course, the resistance changes quite rapidly as the space in the cavity is taken up by the moving lead and expanding jacket.

As soon as the lead and jacket have moved to take up the least possible space inside the die, pressure begins to build rapidly. This is because resistance is increased. No amount of press gain or mechanical advantage will cause any more pressure than the resistance of the material. This is very important to understand. If there is a completely sealed die, with no avenues for the pressure to escape, then eventually all the space in the die will be filled by compressed air and lead and jacket; hopefully there will be a little gap for the air to escape if the die is built correctly. When this happens, then the force applied to the press handle, multiplied by the gain of the leverage and divided by the area of the punch end, will become a pressure that is limited only by the strength of the system to sustain it.

Pressure is force applied to a certain specified area. Another way to say this is pressure is force divided by area. If we apply one pound of force to a rod with an area of one square inch, then the pressure is one pound per square inch. In swaging, we constantly deal with circular area, since the punches are round rods. The area of the end of a round rod is found by the formula:

$$A = \pi r^2, \text{ where } r = \text{radius of the round rod}$$

Another formula using the diameter is:

$$A = 1/4\ \pi d^2 = 0.7854 d^2, \text{ where } d = \text{diameter.}$$

If a .224 caliber bullet is being formed, the punch that seats the lead core in the jacket might be 0.200 inch for a jacket which is about .012 inch thick in the walls at the point where the core will finally rest after seating. This is a very practical example. If the punch is 0.200 inch in diameter, then its area is equal to:

$$0.7854(0.2)^2 = 0.0314 \text{ square inches.}$$

If we were to simply press on the base of this punch by any means, even by putting a finger on it, and the other end rested against a core inside the die with enough resistance to sustain the pressure, any force we apply could be converted into pressure by simply dividing by the area of the punch tip. We could, for example, press the punch against our bathroom scales and read the force directly in pounds. Perhaps we could press fifty pounds against the punch base in this manner. The pressure inside the die, assuming everything came to rest and the lead did not extrude around the punch, would be found by dividing fifty pounds by the area, or 0.0314 square inch. This is 1,592 pounds per square inch. That's a lot of pressure for just pressing by hand, but it is nothing compared to the pressures possible in a loading press.

Note that it does not matter what the size of the punch base might be, nor how many rams or blocks or levers apply that force. The only thing that matters is that the force eventually be channeled to the face of the punch, and that will be the place where the area is measured. The pressure will be constant inside the die from top to bottom provided that the die is closed and there is no pressure escape. During movement of any of the system, there is not a constant pressure but a graduation in pressure through the die. It is during movement, where friction and resistance to flow (viscosity of the fluid, if we think of lead under pressure as a thick fluid) cause a graduation in pressure levels, that those aspects of the final product which are a result of higher or lower pressures are developed. In the point-forming die, for example, there is never really a completely closed system. The top of the die remains unfilled and open, in most cases. Pressure starts at the bottom punch and is greatest against the bottom of the jacket. It causes a slight pressure ring at the base as a result. The straight shank section expands little during this step, and pressure is slightly lower through the shank, dropping only from the effect of friction and die surface finish. Where the ogive starts, there is an abrupt change in the rate of pressure drop. The rate increases greatly, meaning that a good portion of the work being done and energy being consumed is at the start of the ogive. Toward the tip, pressure drops to zero. Just ahead of the jacket there can be no pressure since there is no material in that area.

In the core swage die, there are bleed holes or apertures along the sides of the die so that a fixed amount of lead can be held between two punches designed to protrude an exact distance into the die, and any excess lead is forced to extrude through the holes. As could be expected, during movement of the ram the pressure is greatest at the base of the core, next to the moving punch. It is passed along through the core to the other punch. While the core is swelling and filling the volume of the die, the pressure is greatest at the bottom punch end and almost as great at the top punch end with little or no pressure against the die walls in the middle. Ejecting a core that is cut too short to fill the volume of the die will confirm that the ends are swollen to fit the die but the middle is not.

As the core is brought to fill the entire volume of the die, the pressure begins to increase at a rapid rate. As soon as it reaches the point at which lead can begin to extrude through the holes, it will attempt to reach a level plateau regardless of the force applied to the handle. Actually, the pressure rate of change will increase but at a rate that is more gradual than that for the first stages. Increased force results in more rapid extrusion of the lead through the holes until the friction of the lead trying to pass through the given aperture size balances the pressure developed. If more force is applied, it will eventually be possible to reach such high pressures that the die will explode. This can be a very loud crack, although it is not dangerous since there is very little energy even with such high pressures. The cracking of the die instantly relieves all the pressure. A rifle chamber might have nearly the same pressure but still have a great deal of trapped energy behind that pressure, which would not be relieved except by expending itself in a violent manner. Dies can blow up very politely, usually just falling apart or cracking, never throwing parts around the room.

It is the core-seating die which most nearly approaches a completely sealed system. In core seating, there is no open top to a semiblind cavity. There is no series of apertures for lead to bleed through and relieve pressure. The straight cylinder of the die cavity is filled with a jacket, lead, and snugly fitted punches on either end. The pressure is first applied against the ends of the core, where it swells the ends, and then the middle to fit into the jacket. Next, the core pressure is applied evenly along the walls of the jacket, since the bottom is fixed against the punch face and cannot move. Pressure is developed evenly along the walls but there is a rapid graduation toward zero pressure at the junction of the punch face and the core.

Obviously, there is no pressure against the cavity wall where there is no core to apply it. The mouth of the jacket might be longer than the core, as it usually is with rifle bullets of the open-tip variety. Thus no pressure will be applied to the jacket mouth except where the core touches the jacket. We can expect the mouth to measure slightly smaller than the middle of the jacket. We would not expect there to be a pressure ring at the base of the seated core under normal circumstances, since this is a function of the next step, point forming. But we can see that pressure against the die cavity walls would have the effect of making the cavity want to assume a barrel shape, since pressure is zero at both ends of the cavity and high along the relatively short section where the core is being pressed.

In fact, one of the problems with application of extreme pressures is that the cavity will assume a barrel shape, becoming larger in the middle than at either end. The seated core and jacket also take this shape, of course, since they are what makes the die assume it. Upon ejecting the core and jacket, the fat middle part has to be pressed past the constriction at the die mouth, this being the original size of the die. Even a few ten thousandths of an inch is enough compression to make the ejection hard. Burnished marks around the middle of the seated core and jacket assembly are one indication of this problem. Hard ejection is the main indication.

If the die were cut in half, it would be almost impossible to see a swelling of only two or three ten-thousandths of an inch. Yet dies have been received back under warranty that have swelling that can easily be seen, indicating such pressure that a shooter would be very concerned about firing a cartridge in a rifle with a chamber section no larger than the swage die diameter and developing the same level of pressure. We will presently see just how staggering the pressure can be. The levels are so high that steel rods as large as a person's finger have been mushroomed at the base like so much putty, this being done in a normal reloading press (and possibly by an abnormal operator). Quarter-inch flat-ended punches have been driven through quarter-inch-thick steel plate with what looked like massive blows from a sledge, yet required only that the handloader have no regard for his equipment's precision, not that he have greater than usual strength in his arms. In this instance, a Pacific Multi-Power was used. The R.C.B.S. Rockchucker, the C-H Champion, or any of the compound leverage presses could also be abused to produce the same results. When one considers that the Mity Mite press uses only half the stroke so that it can develop double the mechanical gain of a reloading press, one is not amazed at the results of abusing this potent tool.

If a simple lever arm operated a press ram, and the ram had a stroke of four inches (adequate for reloading), then passing the lever in a half-circle would mean the short arm of the lever could travel four inches with the ram by being pivoted two inches from its end. If the long arm of the handle were twenty inches long, then, this would give us a press with a gain of ten. Every pound applied to the long lever arm would be multiplied into ten pounds at the ram. Of course, this only happens because the end of the lever has to travel ten times the distance as the ram. Either the arc can be made longer, or the lever made longer, or both, to gain power with the same stroke. The stroke can be made shorter, as well, and this will increase the power. For swaging, a two-inch stroke is adequate for nearly all bullets. An inch is required to feed the bullet and jacket into the die, and another inch for swaging.

But with our simple example, we can apply 50 pounds of force to the handle without difficulty, and develop 500 pounds of force on the ram. Using the same .224 core-seating die and punch as we pressed against our bathroom scales earlier, the press can now apply ten times the force, which develops ten times the pressure. Where we had 1,592 psi before, we now can generate 15,920 psi in the same die. This is not a quick pulselike application lasting a few milliseconds as with the chamber pressure in a gun, but a constant pressure sustained as long as the handle is held and the lead does not move. The inertia of the die walls does not influence the ability of the die to resist the pressure.

A swaging die is usually between 5/8 inch and 7/8 inch in diameter. Few rifle breeches are ever this small. So far we have only illustrated a simple fixed-ratio leverage press, but when the compound-leverage press comes into play the pressure situation will be even more extreme. The fixed-leverage press, similar to the lower-priced reloading presses now on the market, will produce the same force no matter where the die is adjusted, since the mechanical gain is constant all through the stroke. The pressure will be determined by the resistance of the core and jacket, but not by any particular adjustment of the die in the press head.

The compound-leverage press is an entirely different story. The main pivot for the leverage is itself free to move about a pivot located on the press frame. The handle runs the ram directly through a pin, and is pinned in its turn to a pair of links that are pinned to the press frame. As the ram moves up, the links swing closer to the axis of the ram, so that the effective ratio of leverage changes constantly and approaches infinity as the short lever arm approaches zero. When the link pivots and handle pivots line up parallel to the axis of the ram, there is an effective zero-length short arm, and all movement stops or approaches a stop as a limit for ram travel, while the handle is still moving.

Thus, the compound-leverage press does not have a fixed mechanical gain, and the amount of gain or force multiplication will depend on where in the stroke the resistance is met. If the die is set so that the punch will come up firmly against the core toward the start of the stroke, then only a simple lower leverage ratio will be in effect. If the die is set so that just as the top end of the stroke is approached, and the links nearly form a straight line, the punch begins to meet firm resistance from the core and jacket, then pressure can be almost unlimited. Usually the gain is between 80 and 100 rather than infinite, because the play in the links, stretch of the metal, and wear in the pivots will not permit any higher gain. In fact, when a more perfect press is designed with all the pins ground and hardened, all the holes lapped for a perfect fit, etc., the limit becomes the destruction of one or more parts in the system. Usually the die explodes, the punch mushes into a wad of putty, or the press cracks apart along a stress riser. How much "feel" does this require? It depends on the adjustment of the die and the perfection of fit in the press. Usually, throwing one's weight on the handle will do the job with a high-precision press.

It is important to realize that pressure in the compound press is limited only by resistance at the point at which the gain approaches infinity. Effort on the press handle may be very slight and still pressures can soar. The reason they do not always reach destructive levels is that there is not enough resistance; the bullet will form just a bit more and take up less space in the die, in a point-forming die. In a core-swaging die, there will be more lead extruded. Only in the core-seating die is there extreme danger of damage to the system by relying on a fixed setting rather than developing a feel for the correct force. A fixed setting is normally good if the jacket and core are perfectly the same each time. But if there is a slightly heavier core, a thicker jacket, perhaps even some foreign material to take up a little extra space, then the fixed setting that developed just a nice level of pressure can push the pressures beyond reason.

For this reason, the core swage and point-forming operations are usually done by adjusting the punch holder or the die (depending on the system) so that the ram goes as far forward as it can just as the desired core weight or the desired tip closure on the bullet occurs. Sometimes a screw may be put into the tabletop below the handle, so that the handle comes up against this. It is very bad to try to put a stop pin against the ram or the leverage, since the pin must sustain the same force as the ram supplies. It is always best to put a stop physically located against the input side of the lever system, on the handle, where force is the least and motion is the greatest. The power is the force times the motion, less friction losses, but it is the force that does the damage to the stop or to the part against which it works.

Because the core-seating and point-forming operations can also be done by feel by an experienced operator, positive stops are seldom put on the press. They are normally added by the handloader who wants them, simply by putting the screw under the handle, into the bench edge. A stop can act as a safety device for the handloader who has to worry about children and curious friends damaging his equipment. With the stop adjusted out, the ram can never go all the way forward and the leverage does not approach the infinite gain position. Force is limited considerably. The force is radically greater for the last fraction of an inch of movement, and this is the only point at which dies explode or presses come apart, assuming no cheater bars or groups of helpful handloaders all jump on the handle.

For many years, it was assumed that swaging required brute force and only the most powerful presses would work. In fact, it is still common for beginners to suggest a home-built hydraulic press or some other high-force system modified to take regular dies. The idea is usually that if a little is good, a lot is better. It is a wonder that the same logic isn't applied to chamber pressure; perhaps some people do this too. The fact is that for the kind of bullets that seem to shoot best and fit the materials we have to work with, meaning drawn jackets and soft lead cores, a compound leverage press with from two to four inches of ram travel and an input leverage gain of from eight to ten on the handle lever (handle of eight to ten times the length of the short lever arm, which is in turn about half the stroke length) is not only adequate for perfect bullets but strong enough to damage the press or dies if used incorrectly. More force is only inviting trouble. It is only usable to the point that the core and jacket fill the volume of the die completely. After that, damage starts.

In core swaging, one can feel the core being compressed and the extrusion start simply because the pressure on the hand (force it takes to move the handle) will increase slightly. In core seating, one can feel the amount of handle travel as compared to the pressure on the hand, first as the jacket expands with the core and next as the jacket touches the die walls and pressure begins to build quickly. If the handle keeps moving slightly under high pressure against the hand, one can even feel the lead extruding past the core-seating punch. Some slight gap is desirable so air can get out of the die, so a punch should not fit so firmly that it scores the jacket wall. This means that with enough pressure the lead core can begin to extrude a fine flange of lead past the punch. The thickness should be like foil, perhaps a half-thousandth to a thousandth of an inch thick at most. Greater thickness only shows that the core and jacket and punch are not well matched.

Air and lead can both go through almost the same size holes with enough pressure, and it is necessary to get the air out of the die if the core is to fill all the volume. This means that a little bit of tolerance is not only permitted but necessary between the core-seating punch and the jacket inside diameter. If a properly made punch is used for the jacket and core weight, and lead still extrudes, then this means too much pressure is being developed. A good test is to determine if the jacket has been expanded into a cylinder that is even from top to bottom, and is larger than it went in. All jackets should be smaller than the die in which they will be used next. Jackets should never be as large or larger than the bore of the next die. A pure lead core without a jacket, swaged in a core swage die, or a core seat die, will come out just a fraction of a thousandth of an inch larger than the jacketed core. The difference is in the amount of elasticity of the copper alloy.

If a pure lead core swaged in a core-seating die does not come out as large or slightly larger than the jacketed core, it means that there is pressure escaping around the punch ends. Naturally, the bottom punch or the one that usually fits into the jacket mouth can't be used to seat a lead core without a jacket, since a punch that fills the entire die bore will be required. But the top punch will show lead extrusion as a fin on the top of the core, if there is much of a gap. The jacket normally obturates or seals this gap, and it is not a defect, but can explain a reason for the lead core not coming out according to the book, larger than the jacket. Lead cores are often used to develop a feel, and to set the dies, since they are less likely to stick or cause any problem in learning. They are also cheaper than jackets and can be used over and over again. Setting the point-forming die correctly with a lead core only is a common procedure. Fine adjustment can be made with the jacket and core later.

As a general rule, the force on the handle will be about like resizing a rifle case full length. This will depend entirely on where the core and jacket are in relation to the possible ram travel, getting less as the ram moves closer to the forward position. In a few millimeters of adjustment, or ram travel, the amount of force felt on the handle can go from nearly impossible to comfortably apply with the hand alone, to a very mild but firm feeling quite easily handled by anyone. Generally, the force would just be enough to put some strain on short wood screws or start to tip a light wooden table or bench, if everything is adjusted correctly. The press need not be welded to a steel plate, but it should be held down by at least a 3/16-inch bolt through the mounting holes, or a long number 10 screw into fairly solid wood. The bench should be just heavy enough or mounted solidly enough so that putting fifty pounds on a two-foot lever bolted to the bench top would not tip it. That is not a particularly stringent requirement. Any moderately solid bench will handle the swaging presses.

In drawing jackets, especially from rimfire cases, we find a very important and radical use of the position of the die compared to the pressure required. The jacket requires a fairly long stroke to push it through a ring die full length. But a rimfire case needs the greatest force just at the start of the stroke, where the head is to be drawn into the body. The contradicting requirements of stroke and force mean a rather fine compromise has to be reached. A little too much stroke being used results in a lot of extra work to push the head into the ring die and draw it. A little too short a stroke, with the die set too high, means very easy operation but only a part of the head is drawn.

A good way to handle this is to draw part of the case with a high setting, and then after having partially drawn a number of cases, set the die lower and run them all the way through. This will distribute the necessary force into two separate steps, and eliminate much of the work required in the first step. Another way is to push the case partly into the die with the die set high, using the last part of the stroke, and then pull the punch down, set a short rod on top of it, and use this longer combination to push the jacket all the way through the die. This gives high force to draw the head and long stroke to push the case through and turn it into a jacket, in two steps. A basic rule of physics is that with a given amount of available power, we can only have more movement or more force, but not both at the same time. No mechanical device will create power for us; we have to supply that.

Chapter 16

Basic Bullet Design

In order to receive the most benefit from custom-made bullet swaging dies, or to use standard designs for innovative kinds of bullets by special techniques of swaging, one should have a basic grasp of the fundamentals for designing practical, stable bullets. Many of the requests for special bullet designs in swaging dies which are received by a custom diemaker are simply not possible. However, unless the customer requests an explanation, he may feel that his design is rejected without good cause. Understanding the limits of die design and of the various components of the bullet would, on the other hand, make such impossible or contradicting specifications obvious.

A good example is the request for a die that will make a bullet with far too sharp an ogive for the weight requested. There is simply no possible way for a .457 caliber rifle bullet weighing 200 grains to have a full 6 caliber ogive in a spitzer design. The basic shank of the bullet would be reduced to less than bore diameter, since the amount of lead required to fill just the ogive part of the bullet would consume the entire 200 grains. No matter how nice such a bullet looks when drawn out, the drawing weighs little compared to reality, and putting a number on it does not mean that it is actually possible to make a real bullet with that weight in the same shape and caliber.

A less extreme example is the frequently requested 8 or 9 caliber ogive bullet shape ending in a very small open tip, usually somewhere around 0.050 inch or less, which seems to be an obsession with some precision long-range shooters. This shape is fine provided that the bullet weighs in toward the heavy side for the caliber, which gives the proper room for a shank at least half the total bullet length. But quite often the request will be for a standard 65- or 75-grain .243 caliber bullet with this shape and tip size. All one has to do is try this once to see that the bullet is nearly all ogive, with little or no shank remaining, certainly not enough for good stability. The advantages of a large-caliber ogive in bucking air resistance have to be matched and balanced against the rest of the bullet design.

One way a person can simulate bullet design without any math involved is to simply make a sample lead bullet to the desired shape and then weigh it. Even without the jacket, it will give a good idea of what weight is possible with any given shape in a certain caliber. A lead core or slug can simply be cast into a plaster or aluminum mold made by turning a wood or steel plug in the desired shape and casting the mold around it. Or, more simply, a lead core can be turned to shape in a lathe or even by hand using a sanding wheel and spinning the lead rod in a drill motor. There are many other ways to make a model lead bullet with the desired ogive, or to simulate the weight of a model turned out of any other material.

A simple simulation is to turn out a sample bullet made of any desired waterproof material like steel, plastic, aluminum, or even wood that is carefully sealed. This model bullet is put into a graduate cylinder with a precise volume of water measured into it. The volume of water displaced is measured on the side of the graduate (the wood model would have to be completely submerged, of course), and then this volume is multiplied by the density of lead to obtain the approximate weight of the bullet itself. One can carry the math a step further, eliminate the model, simply making a drawing to scale, and use this model drawing to calculate by geometric methods the volume of the bullet, which is then multiplied by the density of lead. Or, by more complex higher math the volume and weight can be calculated from the formula for the shape of bullet.

The idea of determining the weight of the particular model bullet is to see if it is really what is desired. It is too easy to simply put figures on a drawing and assume that bullets from "200 to 500 grains" should be made in a certain set of .457 caliber dies, or a similar range in other calibers. It is another thing to actually try to make these bullets, or to see them perform. While the dies are usually capable of handling the job, in theory, the objects they produce may not be bullets. They may be very light cones! On the other end of the spectrum, it might be that a blunt nose very suitable for a lighter bullet, even required for it, would look far different in proportion to the long shank of the heavier weights than the customer desired. Thus a set of dies is usually limited to making practical bullets having about the normal range of weights for a certain ogive that one would see in any factory bullet. The smaller the ogive (blunter the nose), the greater the possible range of weights that still will be stable, potentially, in the rifle. This is not to say that any rifling twist can be used, but only that the shank will be able to support the nose in its trip down the bore, and to do this about half the bullet should be shank and the rest, nose.

The same problem with weight versus shape occurs in handgun bullets which have a step junction and an all-lead nose (three-quarter-jacket style). Here, the minimum weight is that which will leave the entire jacket filled with lead and still have enough for proper filling out of the desired nose cavity in the nose-forming punch. It is not possible to make this bullet style with the weight of lead being less than that required to fill up the jacket. If the jacket is longer than the core, the punch edges will strike it and fold it over. Or, more likely, the jacket will buckle in an irregular manner. Either a shorter jacket must be used with lighter bullets, or the nose must be shortened and made hollow-pointed, or the base must be made with a dish or cup shape to cut down the amount of lead used.

In fact, to maintain the same bearing surface for good alignment, as with small caliber self-loading pistol bullets (many of which are woefully short in regard to the caliber because the weight required is small if the bullet is to have any velocity and yet not produce too much chamber pressure), several tricks of design are worth knowing. First, the nose can be made with a steeper slope in a truncated conical shape, like the Keith styles for revolvers. A spire always uses less lead than a spitzer, which begins and ends at the same points. A round-nose bullet which has a truncated cone shape drawn within its outlines will always require more weight than the truncated cone bullet, if all other features are the same.

Second, a hollow point can be used. This cavity can be made larger inside than at the opening to gain even more in the reducing of weight. This is done by swaging a large cup into the bullet in a core-seating die and then using a point-forming or ogive-walled die to put the nose on this bullet. The cavity will become pear-shaped rather than closing in evenly from top to bottom. The depth of hollow point should never be so great that only a thin web is left between the base and the bottom of the cavity. It might be possible for the base to blow through entirely against the resistance of the jacket in the bore, leaving a jacket and possibly most of the lead in the barrel for the next shot to find.

Third, a cup or hollow base can be used. This is a good method when combined with the other two. The resulting bullet is very light for its bearing surface and ogive, and has better stability in the barrel, and in flight, than the conventional round-nosed bullet. It can also be made lighter than the conventional bullet and driven to higher speeds without the problems of instability this would necessitate in the round-nose design. Both accuracy and increased power or at least a flatter trajectory can be gained at the same time in many cases. A typical example would be the .25 ACP, the .32 ACP, and the .380 ACP bullets which are used today compared to a properly swaged bullet designed for accuracy and stability. Feeding reliability is usually not as good with the special bullets, but this depends on the particular gun and what the handloader does about bullet-seating depth, magazine lip angles, and feed ramp condition and angle. Some automatic pistols will digest very flat-nosed bullets well if the seating depth is correct, while others cannot tolerate deviations from a rather rounded nose.

Because it is desirable to have a bullet with about half shank and half nose or ogive, in most cases, a safe starting point for any caliber of rifle bullet is the 6 caliber ogive. This particular degree of curvature seems to provide the most versatile weights for nearly any caliber in a

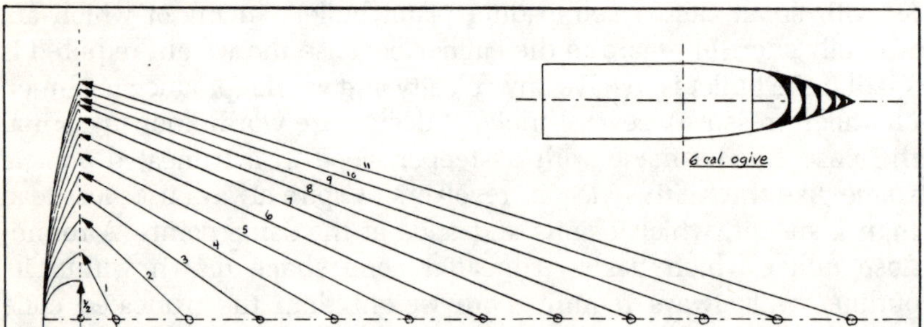

Calibers of radius for ½ to 11 caliber ogives.

spitzer bullet, catching the middle of the weight ranges in everything from .224 to 8mm. Generally, there are few long-radius ogive designs in calibers over 8mm because the purpose to which such bullets are put dictates a deep-penetrating, nontumbling blunt-nose bullet. If a large caliber rifle is to be used for long-range shooting, then a 6 caliber ogive spitzer will still provide reasonable weight and shape even in the .458 bore. Such bullets will usually have to weigh more than their blunt-nosed counterparts, in order to have a good ratio of shank to ogive.

In the handgun calibers, a 2 to 3 caliber ogive will provide the most suitable average in all calibers as a ratio between shank and ogive. Here considerable latitude exists because spitzer shapes are seldom used. The 2 to 3 caliber ogive is meant only as a rough guide, such that the nose shape selected would fit within the outline and touch it at all major changes of direction. A truncated cone, for example, might be drawn so that it came as close as possible to filling the outline of a 2 caliber ogive round-nose shape. It would be found to be quite similar to the usual design found today in the Keith-style bullets.

It can be shown that an angle of approximately 10° (from 9.15° in the .44 magnum to 10° in the .357 magnum) from the bullet centerline will provide a suitable ratio of nose to shank in a reasonable and normal range of bullet weights for revolvers, using the Keith nose form. This is

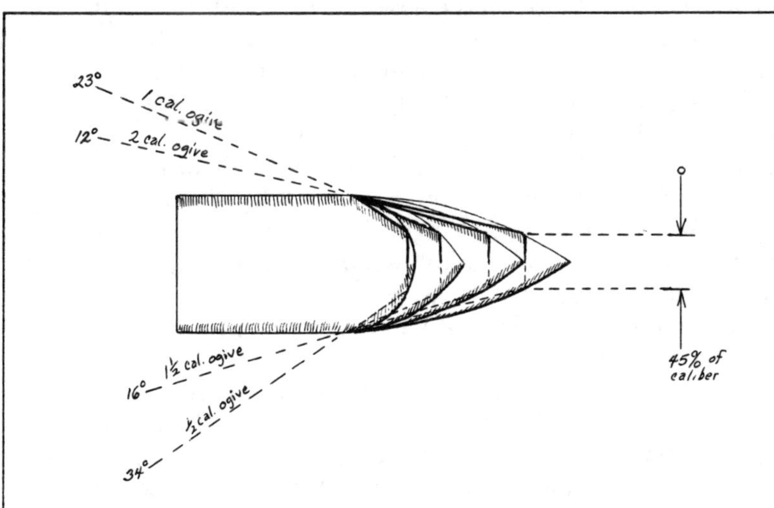

Simulations of ogives in handgun bullets with truncated cone nose designs.

a longer ogive than the 2 caliber, and is more suitable for revolvers. To simulate a 2 caliber ogive in the Keith-style bullet, for use in autoloaders, an angle of 19.5° from the bullet centerline can be used for nearly all calibers. This includes the .25 ACP and the .45 ACP.

Formulas for finding the various dimensions of a bullet nose in the truncated cone style are given in the appendix of this book. The same mathematics are required to design and manufacture the reamers to produce these swaging punches. As with many of the factors in bullet design, there is no absolute rule that a bullet must have these dimensions to perform or feed well. Most of the factors which determine the bullet shape are subjective and have been determined only to an approximation, in an average number of firearms, and although the design guides will produce suitable bullets that appear well balanced, a good part of their appeal is merely visual. Starting with the guidelines, a handloader can go as far in any direction as desired to try to improve on the performance in a certain field. Usually, the further from the guidelines or averages the resulting bullet appears, the less suitable it becomes for all other fields of performance. Eventually, one or more of the other factors is minimized so much that the bullet cannot perform at all.

The continual lengthening of the radius of the ogive to gain better ballistic coefficient is a good example, where the bullet eventually winds up being entirely ogive and will not even fit the bore. Other examples to avoid duplicating would include the specification of too light or too heavy a bullet for the nose shape, specification of too deep a cup or hollow point for the length of bullet, a requirement that a bullet have a longer boattail section than is possible for the weight, or a requirement that a certain jacket be used when the specified weight would preclude its use. A jacket can be used over quite a range of weights, but in order to accomplish this, the style of bullet must be changed after each limit point is reached for the former design. Starting with a minimum weight for a certain length of jacket, open-tip bullets are made. As the weight is increased, eventually the jacket is filled completely.

This is the limit of an open-tip bullet weight for a given jacket. Then the style must change to lead tip. This requires a different core-seating punch, and may also require a different procedure to finish the tip of the bullet, if a sharp spitzer tip is desired. After a series of weight increases in the lead-tip style, a point is reached where the lead will begin to flow back upon firing at higher velocities. Then weight can only continue to increase if the velocity range is reduced considerably. After a certain limit, the ogive is nearly all formed of lead.

A point is reached where any additional weight would require the lead to begin to move into the rifling engagement portion of the ogive. Velocity must be severely reduced to prevent serious leading.

This illustrates that while it is possible to press a jacket into use over an extremely wide range of weights by changing styles of bullet and velocity, it is more practical to use several lengths of jacket when the range of weights is so extreme. In the small bores, a ten- to twenty-grain change in a given jacket is usually possible by going from open to lead tip. In the larger bore, the same approximate ratio would give a twenty-five- to fifty-grain change in the bullet weight. There is no exact bullet weight for a given length of jacket, since the ogive and bullet style will determine the proper weight. A jacket is usually either listed by its length alone, or a certain range of bullet weights is given for it. Unless the bullet style is given, this may be assumed to be an open-tip bullet. Heavier bullets can be made in lead-tip style from the same jacket.

Since lead-tip bullets are more frequently desired for hunting, and open-tip bullets are more suitable for match shooting, the handloader designing bullets which will have to be made in a single set of dies will want to select the style most likely to be used by himself. It is possible to use a spitzer type die to produce round-nosed bullets, by using a tip-forming die and reshaping the exposed lead after the ogive is only partly formed in the point-forming die. But it is not possible to make a spitzer bullet in a round-nose die. Most people, therefore, find that the standard 6 caliber spitzer point-forming die will produce the widest range of bullets for their use. If lead-tip bullets are desired, an additional die to reshape the lead to proper form can be added. The standard set of dies consisting of core swage, core seat, and point-form dies is best suited to open-tip bullet forming, and the fourth die (tip form) is always a good addition for making lead-tip bullets.

The spire point is a good shape except that it does not lend itself as well to a wide range of weights as the spitzer. The spire should have a slightly better ballistic coefficient for making a bullet otherwise identical to the spitzer, since the spire has a slightly greater length for the same weight. The difference is minute, and both the pressure of forcing the jacket to turn the sharper angle between ogive and shank, and the limitations on the partial use of the ogive, such as with making round-nose bullets, usually mean that a spire is less desirable for a swage die set than a spitzer. If a special purpose is planned requiring a spire, this is one thing, but if general experimental use over a wide range of weights and styles is planned, the spire is not the best choice.

If boattail bullets are desired, it is best to consider the extra time and cost of such equipment compared to the benefits of the boattail design. Except for the longer ranges where the bullet velocity is dropping below the speed of sound, the effect of drag at the base is not as noticeable as other ballistic effects. The ease and low cost with which a flat-base bullet can be formed make the boattail design a marginal one for general home swaging. The jacket is formed to a boattail, and all the punches used to seat and form that bullet must have a boattail shape in them. The thin edges of such punches mean they are much more subject to wear than a flat-base punch.

Since there are several extra parts or special steps in manufacturing the special parts, a boattail set of dies is more expensive than flat-base dies. The same dies are used, with the addition of a jacket-preforming die set and the boattail punches. There is the extra step of forming the boattail in the jacket, which is not difficult but is an extra handling of the jacket. These factors should be considered against the possible advantages of a slightly better ballistic coefficient. The effects on accuracy are very much subject to question. Some high-precision bullets use boattails and win national matches, and others win just as often with flat bases. At normal ranges, there is probably very little benefit. It is best to consider the boattail die set only if there is a definite need or desire for this kind of bullet that outweighs the extra cost and steps to make it.

In general, it is best when designing a bullet to leave as much leeway for tolerances as the situation will permit—not in the bore size, particularly, but in the factors such as ogive shape, taper, weight range, etc. If a person desires a set of dies to make a varmint bullet which will be shot in any of several rifles he owns, it seems a great waste of time and money to order dies made to such specifications as a "7.5 caliber ogive" and a "taper not to exceed 0.0002 inch over the entire shank" or even a caliber tolerance of plus or minus "0.0001" at the pressure ring. It is highly unlikely that any two rifles made will be anywhere that close in the best fit of bullet to their bores. Very likely anything from a 6 to 7 caliber ogive, a taper of anywhere from a thousandth of an inch down, and a diameter tolerance of half a thousandth of an inch at the pressure ring would group just as well—or as poorly—as the bullets from the highly specified but also highly arbitrary die set.

On the other hand, if a research program requires extremely high-precision projectiles for a known diameter of bore, then the custom bullet swage set is the best way to obtain such bullets at low cost. It is fashionable to speak in terms of the last ten thousandth of an inch in regard to precision shooting these days, but few who use these

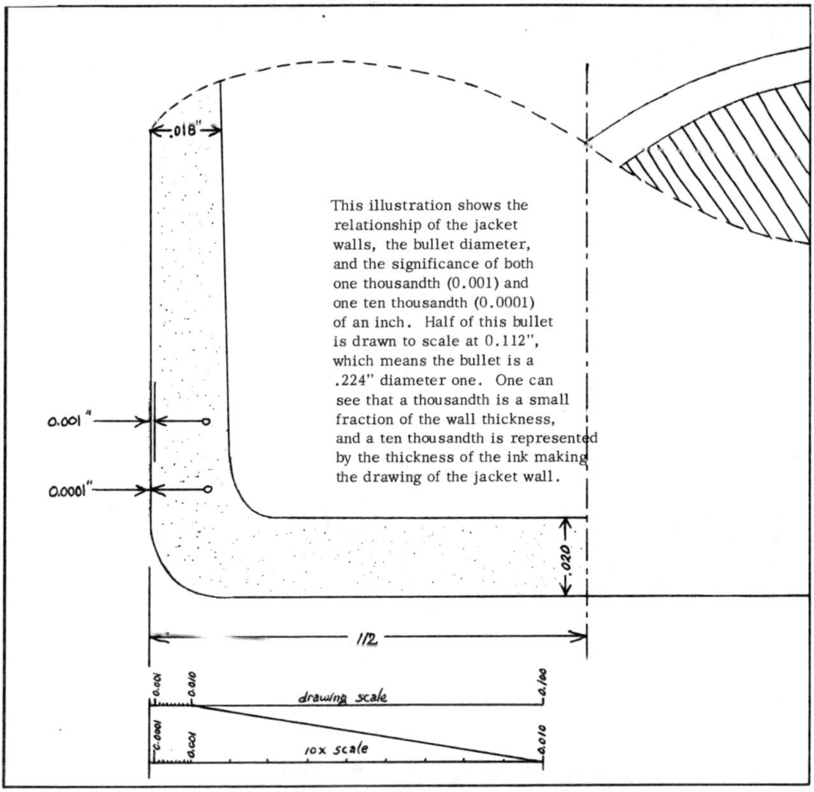

Scale drawing of jacket wall, illustrating tolerances used in swaging work.

terms realize the implications of their statements. Generally, a piece of metal hand-held for a few seconds will warm up enough to expand a ten thousandth of an inch. If such precision is really to be meaningful, it must also specify the temperature of measurement and precise point of measurement. Equipment used outside the controlled, clean atmosphere of a laboratory hardly keeps tolerances of less than a ten thousandth for long.

Chapter 17

Hollow-Point Bullets

In discussing hollow-point bullets, one must draw a distinction between open-tip full metal jacket styles and lead-tip hollow points. The simplest way to distinguish these styles is to say that any bullet with the jacket extending past the core is an open-tip bullet. Any bullet with the core flush with or extending past the jacket, and having a cavity in the nose, is a hollow-point. The difference is important, both in terms of forming the bullets and in how they will perform.

If a core is seated within the jacket, using a punch that fits the inside diameter of the jacket, and then this seated core is formed into a bullet in a point-forming die, either of the kinds of bullets may be made. Usually, the lead will move forward but not sufficiently to extrude past the jacket tip. This will leave a sort of windshield at the tip, made of empty jacket metal. The ballistic form will be preserved but the weight will be a little less than a completely filled jacket of the same form would require. Somewhat higher speeds may be obtained with this bullet, therefore, but the main reason this open-tip style is used so much for match shooting is the superior accuracy.

Accuracy can be equally good with a lead-tip bullet, but the chances for damage to the tip are greater. At very high speeds the tip is deformed against the shock wave itself, so that the attractive sharp point of a lead-tip spitzer may be good only for the time the bullet remains at rest! At higher speeds, even the relatively flat tip of the hollow-point lead-tip bullet can flow back against the pressure of the atmosphere, resulting in fliers and unpredictable performance. However, the open-tip bullet does not expand nearly as well as the hollow-point, or for that matter a solid lead-tip bullet, given the same speed and jacket material. This is obvious from testing and the reason is just as simple to see: the jacket is the factor most responsible for controlling expansion. If the jacket is closed down to a small open tip, then the ease with which this mass of copper and zinc alloy can be peeled back is much less than if it is struck against a target while being nearly an open tube. In a lead-tip bullet, the diameter of the jacket opening is much greater than the open-tip style. Thus it follows that since the lead itself has little to do with expansion control, the lead-tip bullet can and will open up more easily and quickly than the open-tip.

The hollow point can be formed in many ways depending on the velocity and target material against which it will be used. A very deep hollow point with thin walls offers faster expansion, but will also shear away and fragment more easily at higher than optimum speed. A shallow cup point offers strength and good opening rate at higher velocities, as well as more sudden deceleration upon impact (which would be related to increasing the shock effect). But it may fail to expand much at very low speeds, and is not usually as good a ballistic form for long-range shooting.

Attempts have been made over the years to combine both effects, so that a bullet might strike anywhere in its effective range and offer good expansion, yet remain essentially in one mass to allow deeper penetration and maximum shock effect. The results are usually not successful at all ranges, or against all types of targets. The range of force which will act on the bullet at various impact speeds, and against different kinds of target material, is simply too wide to design one effective mechanism offering reliable expansion. It remains up to the shooter to select proper range and target for the velocity and bullet that will be used.

Still, there are merits to the various designs which help broaden the range in which a bullet can perform. It is helpful to know the various ways to form hollow-point bullets for both rifle and handgun use. By testing against the actual targets under different conditions over a period of time, a range of velocity may be determined within which the projectiles operate properly. One of the most simple ways to form a hollow-point bullet is to seat the core with a punch that has a conical projection on it.

The core must be longer than the jacket if it is to form a lead tip, and the punch used to seat it must fit the core-seating die, rather than the inside of the jacket used. There is a transition range in weights of cores for any given jacket, in which the lead is longer than the jacket before seating, but is seated so it swells and shortens enough to be within the jacket. Such cores may be seated with a punch that is a good fit to the jacket mouth, but only if the punch is very carefully aligned on the jacket after the lead is compressed almost even with the jacket. It is better to seat it very carefully with a full die-bore diameter punch and then switch to a smaller punch which fits the jacket—or, since the difference in weight is extremely small between this transition range and the true open-tip or lead-tip lengths of core, avoid this range entirely.

If a punch smaller than the jacket mouth is used to seat the core

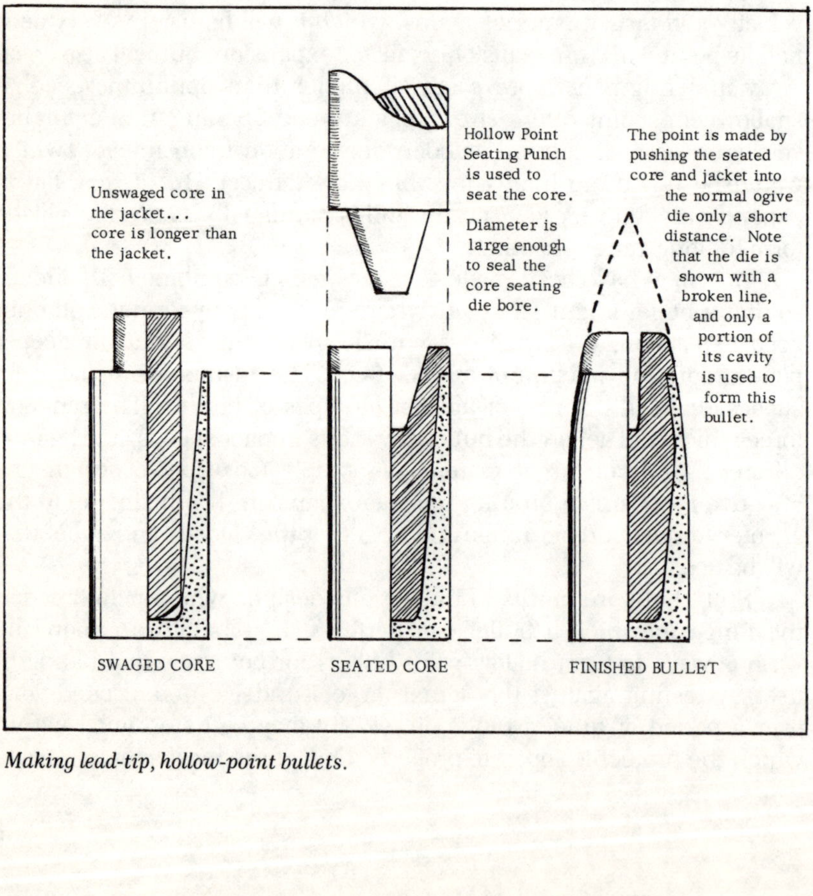

Making lead-tip, hollow-point bullets.

longer than the jacket, lead will simply extrude around the punch, between it and the die walls. If a full die-bore size punch is used and the core is shorter than the jacket after seating, the end of the jacket will contact the punch tip and the force will crush the jacket. Neither of these situations harms the die, but no usable bullets result.

When a point-forming die is to be used following a hollow-pointed, seated-core operation, the shape of the cavity will be changed from that of the hollow-pointing punch. The force that brings the jacket down into a smooth ogive curve also pushes the cavity walls of the hollow point together. The smaller, narrow hollow points can be swaged completely out when the point is formed. Usually, a much wider and deeper cavity is required on the seated core than will result after a significant amount of ogive is formed.

On handgun bullets, where the jacket may just barely be turned away from full bore diameter in the point-forming die, a shallow cuplike cavity can result in a more normal-looking hollow point. Usually it is best not to form up the ogive on the bullet to the full extent possible in a given die, if a nice hollow point is going to be retained. Simply adjust the die a little higher in the standard reloading press, or adjust the punch further back (less penetration into the die) in special swaging presses with dies fixed in position in the ram. Since there is no sharp tip on a hollow-point bullet, there is no need to use that part of the die cavity which has the final tip shape in it.

The end of the bullet, across the flat of the tip, is not supported by any die walls when the bullet is formed in the point-forming die. The lead may turn and extrude in an irregular manner, and thus give a somewhat rough-looking tip. Only those parts of the bullet which have been pressed against a smooth die or punch surface will acquire the same smooth surface. To finish the bullet for the sake of appearance (no practical effect on the accuracy or performance results), it may be ejected, reversed, and tapped very gently with the bottom punch while in the point-forming die. Only a very light pressure is needed, and the base will not acquire an ogive as a result. (This is a good way to put a rounded or even slightly boattailed base on a bullet, providing the bullet nose will stand the pressure and not deform excessively.)

A simple trick to finish off the tip of a bullet of either open-tip or hollow-point style is to drop a proper size lead shot into either the open jacket or the cavity of the hollow point before the point is formed. Sometimes, this may work best if the point is partly formed up and a small lead shot dropped into place. Then the rest of the ogive is formed. The idea is to close off the end of the cavity, or to fill the tip of the jacket, with an alloy harder than the core lead itself. The standard chilled shot is hard enough to resist deformation in handling, has a good shape if the proper size is selected, and since it is not actually swaged, it does not contribute to higher pressures or touch the die walls.

On impact, the lead tip with the harder lead ball in the top of the cavity is supposed to be opened from the ball driving back into it. In practice, this may not occur every time depending on the velocity and how quickly the bullet is decelerated. A hard steel ball can be used instead, but there is no real advantage and it is possible to damage the swaging die if too much pressure is applied. At one time, a company modified factory ammunition by drilling a similar cavity in the points of bullets and inserting a ball in just this manner. The company claimed rather impressive performance for the bullets, but like all mechanical aids to expansion, it was not completely dependable. The modified bullets were called "Hoxie" bullets.

Unusual effects may be obtained by pressing a hollow-point punch into the seated core after it has been seated with a flat punch in the usual manner, or by following a cavity type handgun nose punch with a hollow-point punch. The shape is changed because the lead in the cavity area is displaced. A person owning a Keith type handgun nose punch can thus obtain several other shapes of noses by pressing a single shape of hollow-point punch into the bullets. The degree of penetration will control how much lead is displaced, and thus how much the outside shape of the nose is modified. A rounder nose results from putting a cavity into a Keith nose bullet this way.

In a similar manner, the hollow point may be formed in a full wadcutter slug first, and then the various shapes of handgun nose punches used to swage a new nose. The shape will be somewhat different from that obtained by using a nose punch alone, unless enough force is used to completely re-form the slug. Various degrees of re-forming are possible by using lighter than normal pressure, and thus different amounts of punch penetration into the die. In nearly all cases, a more finished bullet results from gently bumping the nose against a flat punch after the hollow point is formed.

The mystery of the pear-shaped hollow cavity is of course nothing more than the result of forming up an ogive on a slug with a rather wide and deep cavity preformed into it during core seating. The end of the bullet is greatly drawn down, which closes off the end of the cavity, while the shank of the bullet is not reduced. If the cavity extends far enough toward the shank, it will not be closed for its full length. A bullet appearing to be solid-nosed, but having a cavity within the nose, can be made this way. It converts rapidly into a wadcutter style bullet when it strikes, as the cavity collapses and presents a dramatic change in ballistic shape. There is nothing particularly better about this bullet than an ordinary hollow-point, except that the cavity is not obvious. In some police situations this may be an advantage.

A hollow point may be formed even in a jacket, if the punch is stout enough. A full-jacket bullet can be made by swaging a bullet with the base of the jacket toward the ogive of the die (insert seated core backward). The cavity can be formed in the base of the jacket during core seating just as with lead tips. Another way to make a full-jacketed bullet is to put two half-jackets, a half-jacket and a three-quarter jacket, or even a pair of three-quarter jackets end to end over the proper length of core. The assembly makes a fully covered slug when the jacket ends come together. The core weight and thus length is critical so that the jackets do not strike and then fold.

If a hollow-point punch is used to seat one end of this assembly, it will reduce the amount of lead required for the same jacket combination. Using a hollow point in this instance makes it possible to control weights to a finer degree. There is little practical application of the hollow point as far as aiding expansion is concerned; it will of course tend toward opening, but the base of a jacket is usually quite thick compared to the mouth. These two-jacket bullets tend to expand where the junction of the jackets appears, rather than at the tip.

A fine degree of weight control may be achieved by calculating the amount of lead in the cone which the hollow point removes. The formula for the volume of the truncated cone appears in the appendix of this book. When working with very light bullets, such as in experimenting with small autoloading pistols or high-velocity rifles, it is sometimes desirable to obtain the maximum length of bullet for a given weight. Calculating the weight reduction gained by hollow points, hollow bases, or thicker jackets can aid in designing such bullets.

Chapter 18

Boattail Bullets

There are two basic kinds of boattail bullets. The standard boattail has an angled section between the shank and the base, usually between nine and fifteen degrees. The boattail reduces drag at the base of the bullet, which becomes a greater factor below the speed of sound. Bullets fired at low velocity or at longer ranges where velocity drops below the speed of sound are particularly affected by the improvement in ballistics. A boattail can be a problem for the bullet swager because the thin edges necessary on the punches, both to form and to maintain the boattail shape, mean short life and easily damaged punch ends.

A solution that also seems to help make accurate bullets and to reduce the amount of gas cutting of the bullet caused by gas pressure pressing in at the boattail section and seeping past the lower shank portion is the use of a rebated boattail. The rebate is a step, either a sharp angle or a small radius starting at 90 degrees to the shank, that joins the angle of the boattail section to the full bore size of the bullet. This step may assist accuracy by causing a very fast transition between sealing the muzzle gas and releasing the muzzle blast, which is better than a slower release from a gradually increasing clearance between bullet and bore. Any unevenness in the angle of the boattail or wobble of the bullet can be exaggerated by the escape of gas in greater volume on one side of the bullet as it clears the muzzle. A rebated boattail has a sharp step that is in the transition zone much more briefly than the angled seal of a regular boattail.

Also important may be the surface perpendicular to the bore caused by the rebate, which is not compressed inward to allow gas pressure to seep around into the shank portion and thus cut into the bullet at the rear of the seal it makes with the rifling. The sharp step affords a place for metal displaced by rifling to fold back at the rear of the shank, while the boattail has less space for this metal. All these things are probably minor effects if they actually do exist. Important to the bullet swager is the increased strength and longer life of the punches that result from having the step. This step or rebate is the area which represents the punch edge. Unlike a boattail punch, the rebated boattail punch does not have to go to a feather edge to form a smooth junction between shank and boattail angle.

Although many methods have been used to form boattails on bullets, one of the most satisfactory is the use of a preform die for the first core-seating operation, followed by a regular core-seating die and boattail punch. In a normal flat base bullet, the core has a slight dome swaged on one end just to ensure that it will not trap air under it against the bullet jacket bottom. This core is put into the jacket, and a pair of flat-ended punches press on the core, one directly on the core through the jacket mouth, and the other on the jacket base. The core is seated in this manner in a straight cylinder cavity die.

In the boattail system, the core can be swaged in the same die, but one of the punches has a boattail cavity machined into its end. Usually, this punch is the one that rests within the die. With normal core swage punches, the punch can be removed fairly easily after use from the die. The boattail punch has thin walls around the cavity in its end, which allows the punch to spread under pressure. The edges of the punch expand to a tight fit to the die cavity and if the punch can be removed by hand, it will be very difficult to replace it without bending the edges of the punch. For this reason, core swages for boattail bullets should normally be kept assembled with their internal punch, and a separate core swage die used with a flat-base punch rather than changing continually. With care, one can change punches in a single die to go from flat to boattail, but having a second complete die avoids potential damage to the punch edges.

The core that results has a boattail shape on one end. This core is put into a jacket, and the assembly is put into a special preforming die.

The preforming die could be called the first core seater. It is shaped just like the boattail that will be formed, except that it may have a degree or two less taper than the final shape. Also, it may not have a rebate or step, since this is easily added with a punch, and is in fact hard to avoid if the punch ends are very thick. Having the angled portion at one end means that the die is more like a point-forming die than a core-seating die. It has a semiblind hole, and the ejection punch is made so it is just the size of the bullet base. This punch has locking nuts that allow it to be set precisely at the bottom of the cavity, so its flat end forms the bottom of the die cavity and rests exactly against the jacket base.

Pressing the core into the jacket will cause the jacket to take on the shape of the die cavity. A punch with a flat end, exactly the same as if a flat-base bullet were to be formed, is used. This punch only need fit the jacket internal diameter for the length of lead core being used and the brand of jacket. It can also be used with flat-base bullets. The preform die is only used to make boattail bullets. The top punch, or internal punch, ejects the bullet jacket and core, and at this point the jacket is a cylinder with a standard boattail formed on it.

If a standard boattail shape is desired, very thin-edged punches will be used to press the jacket and core still more firmly together in the straight-wall core seater. The punch with the cavity in the shape of the boattail will again be located in the die. Sometimes a slightly smaller diameter of core-seating punch will be used, since the core is now seated deeply and a better seal can be had against the internally tapered jacket walls with a smaller punch. Better seals mean higher pressures on the core with the same effort on the press. This in turn makes sharper turns at the boattail angles and a more professional-looking bullet.

If a rebated boattail is desired, the only difference is a slightly thicker edge on the punch, around the cavity. The punch for rebated boattails does expand and is difficult to remove by hand from the die, as well as being difficult to insert by hand without some pressure being put on the edges of the punch and bending them. The regular boattail shape requires very thin punch edges, and this means almost certain damage in changing back and forth to flat base. Normally, it is better to keep the punch in the die and never remove it. If this requires two dies, one for flat-base and one for cup-base bullets, then it is part of the price one pays for the extra performance. A careful person can get by with one die and a punch for both kinds of bases, but damage is hard to avoid on standard boattail punches and slightly easier to avoid on rebated types.

The core is seated in the normal manner. This same die, with flat-base punch, would be used to make a flat base-bullet. The core and jacket then are formed in the final shape, with the final degree of boattail angle. The slight difference in angle between the preform and core seat number two punch helps assure even forming and prevents sticking. The cardinal principle in swaging is always to swage up, never attempt to draw down in a swage die. A slight increase in angle is similar to swaging upward, in that it allows the jacket elasticity to work for, rather than against, the handloader when it comes time to eject the bullet.

The next step is point forming. The only change here from flat-base bullets is the use of an exact duplicate cavity in the bottom punch or external punch that pushes the bullet into the die. This punch has a short life if it is used carelessly. The bullet shank will guide the punch edges into the die cavity, but care must be used that the edges of the punch never contact the die mouth. If the edges do strike the die, they will be bent and probably will not be straightened out properly under pressure of swaging. This will leave an irregular indentation in the bullet at the rebated section, which is very bad since this is the part that last seals the gas at the muzzle of the gun. Any premature escape of gas can tip the bullet as it flies out the muzzle.

This punch is subject to less swelling than the others, since it is used in an operation with less pressure requirements. The higher the pressure, the more difficult any cavity form punch will be to remove from the die. Since forming the ogive on the jacket requires less pressure than seating the core or swaging the core, it is practical to insert and remove this punch on every stroke, as is required to press the bullet into the die. Naturally, it can be changed to flat base at will with no problem and the same point-forming die used with flat-base seated cores and jackets. The weight of a boattail bullet will be less for the same jacket length, or conversely, if a similar weight is needed in boattail configuration, either a longer jacket or more exposed lead at the tip will be required.

The difference is such that a 150-grain flat-base bullet which required a one-inch-long jacket might then require a 1.125-inch-long jacket for making a boattail bullet, both having the same nose shape and amount of lead exposed at the tip. Or, a 168-grain flat-base bullet might make into a 150-grain boattail bullet with the same set of dies and jackets. One advantage of the boattail that has nothing to do with the drag itself at the base is the extra length for a given weight. Such a bullet may require a slightly faster twist rate to be stable, or it may be fired at a slightly higher speed with the same rifling rate of twist and still be accurate at shorter ranges. The problem with a degree of initial instability from overly fast speeds of a given bullet length, caused by the velocity being a bit too quick for the rifling rate and resulting in a wobble in the bullet's flight path until rotational velocity can be shed, may be solved by going to the longer boattail bullet in the same weight.

The swager who is attempting to decide whether or not he should order boattail dies and punches should consider both the extra expense, the potential for some damage to punches, and the extra step involved. These should be balanced against the gains in flat trajectory (which has to be considered in light of the range and velocity of the bullet, as the gain is not significant for many purposes) and other technical parameters. The effect at normal ranges and speeds is not extreme enough to warrant the cost and time involved, for most hunting and target shooting. It is faster and just as easy to make high-accuracy flat-base bullets for benchrest work or varmint shooting up to perhaps 400 yards. The question of accuracy is not based on which base of bullet is better, but on which will be more consistently swaged and inspected. Either will do as well as the other if both are made correctly.

There is an extra step in forming the boattail bullets. This is the preform or first core seat operation. Since it does require an extra step, it potentially can introduce errors. Whether it does or not simply depends on the care of the operator. If there is no compelling reason to have a boattail bullet, the swager is better advised to obtain flat-base dies and punches. They can be converted to boattail later by adding these items: core swage boattail punch (internal), core seat preforming die and punch set, boattail punch for core seat die (internal), and boattail punch for point-forming die. If a lead-tip bullet is desired and either a radius or spitzer tip on the lead is needed, then a lead tip-forming die is used after point forming. This die will also need a boattail punch (external) for use with boattail bullets.

The maximum number of presses, each set up to perform one operation, would be (1) core swage, (2) preform, (3) core seat, (4) point form, (5) tip form, or a total of five. In the flat-base bullet, this would be simply (1) core swage, (2) core seat, (3) point form, (4) tip form. A boattail can be formed with just a punch, in the standard core-seating operation, but generally this is not as satisfactory a method and it does not lend itself to as nice a boattail. In the past, some attempts have been made to do this and have been moderately successful. One of the biggest problems is wear and damage to the thin edges of the punch that forms the boattail. Great pressure can be developed, which may impress the edge of the punch into the jacket wall and cause a ring where none is desired on the bullet. Forming the boattail against a die cavity rather than a punch end cavity eliminates the need for any junction between bullet shank and start of the boattail.

A sort of boattail effect can be had by simply reversing a bullet in the normal point-forming die and swaging it lightly by pressing backward into the die. This will damage the nose and change the ogive if carried to extremes but it does tend to put a slight ogive on the base of the bullet. For experiments and short runs this is an effective method. On handgun bullets, a slight radius may reduce the damage to case mouths in reloading, or eliminate the need to do any belling of the case mouths. It is probably best to combine this with a cupped base so that gas pressure can move the jacket out to seal the bore rather than having gas pressure working against the angle at the base and compressing the jacket away from the barrel. In a revolver with the change in dimensions between the cylinder and the barrel this can be

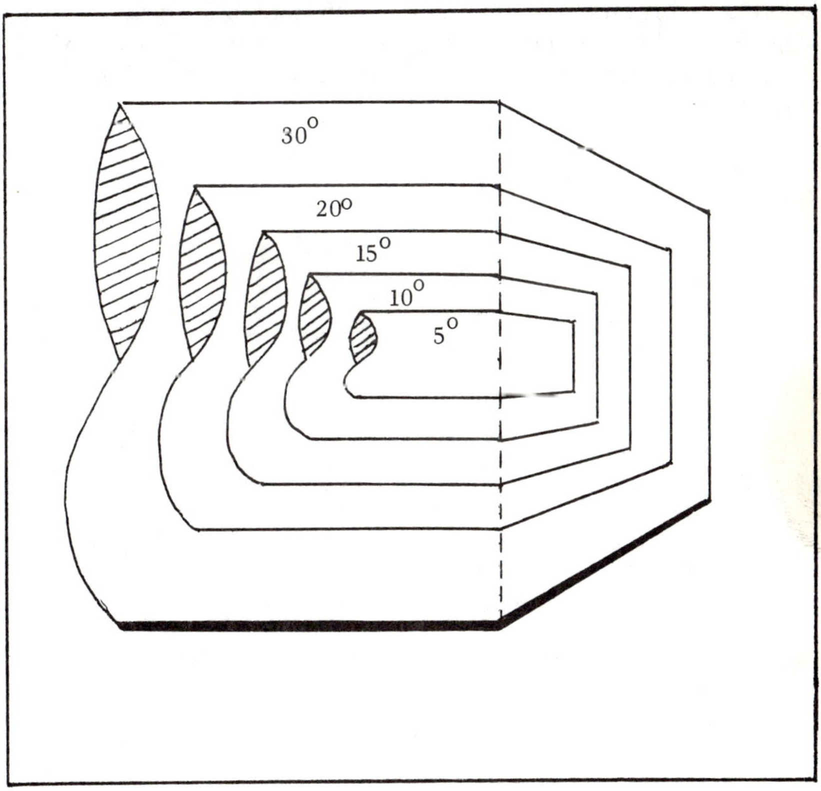

Chart of boattail angles.

a factor in preventing gas cutting of the top strap of the gun's frame. Boattail pistol bullets are not usually reasonable since the length becomes too long for the weight, causing feeding problems in automatics, stability problems in all the standard twist barrels, potential for gas to escape and cut either the frame of the revolver or the bullet jacket, or both, and generally the ranges are such that any potential advantage is lost.

In theory, however, a properly made boattail pistol bullet would make an excellent long-range projectile for such uses as handgun metallic silhouette shooting. This is the realm of the advanced handloader and experimenter, but it is made available for testing by bullet swaging. Such bullets as the double-ended or football-shaped projectile are also possible using the boattail swaging system and special shapes to the preform die and punches. Such bullets have been tried in the past, using sabots to seal the bore. The experimenter should try to become familiar with older work with special projectiles before ordering swaging equipment to duplicate these ideas. German, English, and United States military writings cover many ideas which would appear novel today, though dating back to the turn of the century.

Chapter 19

Copper Tubing Bullets

Sometimes there is no jacket available for a larger-caliber rifle bullet and the swager is faced with finding a substitute. Copper tubing is a readily available choice with certain advantages and disadvantages in performance. The very first jacketed bullets used copper and many special bullets such as high-velocity subcalibers (the .17 is an example) were custom loaded with solid copper bullets. Copper does have less strength to resist twisting across the rifling when used with high-velocity loads than gilding metal alloys. It tends to foul worse than the alloys of copper and zinc, but the fouling is not very serious or difficult to remove.

On the plus side, copper tubing jackets will expand very nicely without cracking apart, and, especially if soldered to the core, they permit some of the very finest performance on big game, with classic mushrooms and high retained weight. Tubing is available in a wider range of thicknesses than most jacket materials. It is fairly low in cost, and not too difficult to cut to length or swage. In order to close the base using nothing but the material of the jacket, considerable force is required along with more length of stroke than the average press can handle. An arbor press is one of the best tools for swaging closed bases on copper tubing jackets.

There are two basic ways to close the base. One is to start a boattail on the jacket by pressing it into a die with the angles polished carefully into the walls. The boattail length is such that when pressure is applied with a flat punch, the boattail section will fold over and make a closed base, with just a pin-hole in the center. This takes a good deal of pressure when the jacket thickness is much over twenty-five-thousandths of an inch. It is not recommended for use in any lever-operated bench press except for arbor type presses, since all the leverage presses develop their power at the end of the stroke. This operation requires full power over the entire stroke length, and only geared or hydraulic presses can develop both stroke and power sufficient to handle the job.

The second way to close the base is to actually swage copper from the inside of the jacket and push it down to the base, forming a new surface at the bottom of the tube from the relocated metal. A punch that actually scrapes out the inside of the jacket is used. The punch is very hard and tough. A great deal of force is needed for this. The punch plows up enough material from the inside of the jacket, as it passes through, to swage into a solid base at the bottom of the tube. Again, only geared (arbor) or hydraulics can handle the stroke and pressure requirements at the same time. A large punch press would be necessary to have enough power, since at least an inch of nearly full-power stroke is required. A punch press, like the lever presses used for handloading, develops power only at the end of the stroke and has less pressure available as the ram moves farther from the bottom of the stroke.

This does not mean that such jackets cannot be made at home, since arbor presses of about five-ton capacity and at least an inch of stroke are not completely beyond the means of most handloaders. But there are easier ways to provide for lead protection at the base. In fact, in many cases there is no need to close the base. Many of the larger-bore bullets will be driven at such low velocities and low pressures that an open lead base would cause no particular harm. All one has to do is consider that nearly all military full-jacket bullets are open at the base, and many of them are used with quite stringent weapons requirements, such as machine guns. If the exposed lead base were going to cause any problems, it would surely foul a hot-running machine gun bore more quickly than a favorite big-bore rifle.

Still a protected base is a good idea and it can be achieved with little trouble by using a smaller caliber gas check or short jacket inside the copper tube jacket. As an example, a .375 caliber bullet can be made out of 3/8-inch refrigeration tubing with thirty-five-thousandths-thick walls, and a .308 caliber jacket used inside the tube for a well-fitting base. This is an example of a multiple-jacket bullet, used to produce jackets when tooling and time is absent for anything more elaborate. The looks and performance are so good that one might well consider if there is any reason to go to further trouble in closing the bases.

With very hot loads and lead-tip bullets, there is a chance that the core might be pushed out of the tubing jacket, leaving the jacket in the bore. This is a remote possibility but to prevent it, the tube and short jacket can be soldered together prior to swaging in the core. In fact, if the insides of the jacket and tube are tinned by brushing with a solder and flux paste mix while applying heat to the outside of the jacket and a clean core is used, it will bond itself to the solder plating, making a completely soldered core that cannot be blown out of the jacket. Such bonding is always done before the bullet is swaged, so that gas and flux will be swaged out and a tighter assembly will result. Heating after swaging can change the dimensions of the bullet and result in gas pockets within the bullet.

In selecting a diameter of tubing to make jackets, one must remember that the tubing has to be smaller than the core-seating die bore, which means it has to be at least a thousandth of an inch under finished bullet diameter, if not several thousandths. It is always best to use a jacket that will be expanded up in core seating. It is very surprising just how much a jacket can be expanded upward during this step and make excellent bullets. A .355-inch jacket can be expanded to as much as .375 inch without difficulty. This permits the use of .38 handgun jackets for making .375 caliber rifle bullets, using a pair of jackets end to end over a lead core. Unfortunately, copper tubing is not always available in just the right diameter for jackets.

To use common sizes of tubing for jackets, it is usually necessary to draw the jackets through a ring die. This will polish them slightly, and remove some of the surface blemishes, while reducing the diameter to precisely that required for the core-seating operation. A 3/8-inch tubing section cut to length for a .375 caliber bullet is usually about an inch long. It is not difficult to press this inch of tubing through a ring die, even with a regular loading press, and reduce it a few thousandths of an inch so it can be swaged back to .375 caliber during the two swaging operations that follow (core seating and point forming).

To reduce the tubing to .358 caliber would require considerably more pressure. This can be done in short strokes on a loading press, by starting with the die set very high in the press and using just the last few millimeters of stroke of the ram to work the tubing through the die. Of course, an arbor press is ideal for this kind of work since higher pressure is required over a long stroke. It is not that such high pressures are required as that the normal pressures developed by a loading press are required over the entire stroke. For the .400 bore rifles, a 7/16-inch tubing diameter is ideal. The thicker the tubing walls, the more difficult it will be to swage up a smaller-diameter jacket. If thinner-walled tubing is used, then even a 3/8-inch diameter tube would suffice. With the usual forty-thousandths-thick walls, swaging up from 3/8-inch to a .404 or .405 bullet size would be difficult in a hand press.

For .458 rifles, the nearest common tubing size is half-inch diameter. This can usually be reduced two or three draws to proper diameter. A single pass through a .455-inch die would be rather difficult and could buckle the jacket rather than smoothly press it down in diameter. When reducing larger calibers by more than a few thousandths of an inch, it is usually necessary to split up the reduction into steps of not greater than fifteen- to twenty-thousandths of an inch. Very thin tubing may be reduced only in small steps because the jacket section is supported only by its own walls at the bottom of the die and punch. All the pressure of drawing is applied by a shoulder on a closely fitted punch that slips inside the jacket to help keep it from buckling. There is little support to the jacket before it passes through the constriction in the draw die, since the punch has to be small enough to fit inside the jacket after it is reduced in diameter. Thus, a thick-walled jacket is strong enough to support the pressure of drawing as it is pushed through the die by its own bottom edge. A thin jacket may fold or buckle unless very gentle pressure is applied, which means small steps of only a very few thousandths of an inch at a time.

The alloy of tubing makes a considerable difference in how much pressure is required and how much reduction can be made at one pass. Ordinary refrigeration tubing is supposed to be 99 percent copper or better. It draws easily but has little strength to support its own pressure of drawing. Alloys containing zinc, such as the various gilding metals and even brass tubing, can usually be drawn in greater steps and in thinner-walled sections. Such alloyed tubes are available on special order from many of the copper and brass suppliers. Some of the brass tubes can be ordered in close-fitting sizes that telescope within each other. Such material is used for moving joints and industrial applications with sliding fits of tubing.

Very thick-walled tubing such as ordinary water piping is generally not suitable for home swaging, since the pressure required to form the point or to expand the jacket in core seating is very high. Special extra-thick dies can be made for arbor presses or hydraulic presses to use water pipe jackets. Dies made for commercial jackets in lever-operated presses are generally too thin in their walls for this kind of pressure. The thicker the jacket wall, the larger the caliber must be so that the ogive can be formed without wrinkles and folds. Very thick jacket walls, such as those over fifty-thousandths of an inch, are best used only with .458 and larger bores. If limited expansion is required, multiple jackets or partitions should be considered rather than swaging very thick jackets in regular swage dies.

The tooling for closing the bases of tubing jackets is relatively

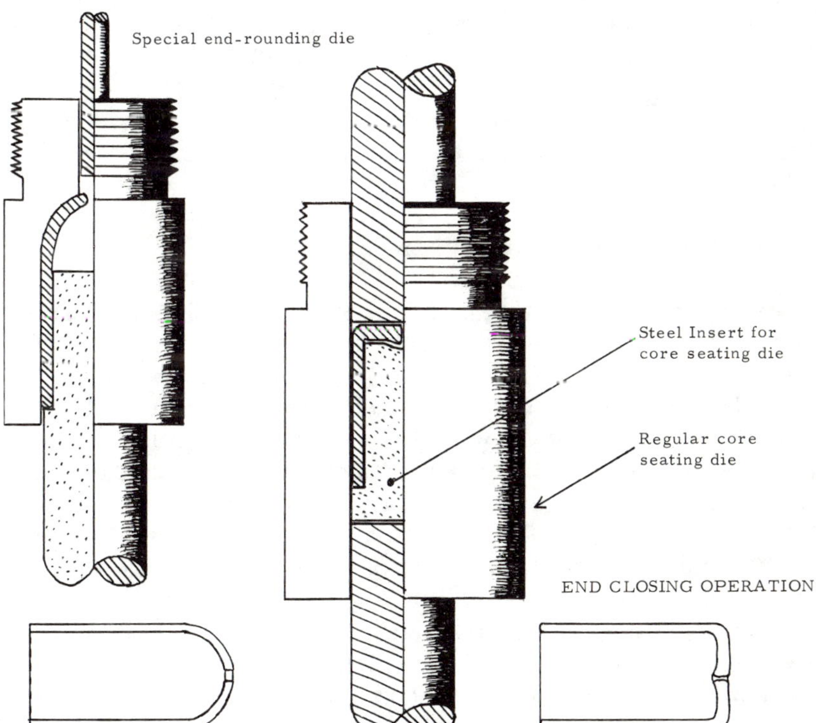

Making jackets from copper tubing.

simple although much more costly than just using gas checks or short jackets for the base. The first step is to put a boattail on the empty jacket. This is done by slipping the jacket tube over a punch with a sharp shoulder, which then presses the empty jacket into a die with the boattail shape lapped into it. A large punch which comes exactly to the bottom of the die from the other end, meeting the boattail edge of the jacket after it is formed, presses the jacket back out again. A core is put into the jacket before the base is closed completely. A flat-base punch is used to seat the core and to press on the bottom of the jacket. During core-seating, the pressure against the base causes it to fold inward and become flat. This is further refined in the point-forming operation.

The process works best with thinner jackets. Thick jackets tend to fold only to make a slight rim around the base, and not to flatten completely toward the center of the bullet. A gas check can be dropped into the bottom of the tube before the core is inserted, and this will help close the base in case the tubing material wants to leave an opening in the middle. The pressure of swaging will usually form the gas check so that it comes flush with the rest of the base, protruding through the opening.

For thick jackets, the best process is the scraping and swaging of surplus material from inside the jacket walls to make a base. The tubing is held in a die while a punch is guided by the die, using a larger-diameter section on the punch as the guide surface. The smaller-diameter section is very hard and has a sharp edge that plows up copper and presses it to the closed bottom of the die. The closure is made by a fixed punch having a flat end. This punch is used as the ejection punch when the base has been finished. Although in theory the die could be inserted into a loading press, and the top punch used as the bottom of the die while the guided hardened punch was fitted to the ram, there is not enough pressure over the entire stroke to accomplish this. A five-ton arbor press would be able to handle the job if it had at least a one-inch stroke for power and a one-inch stroke in addition for loading the tube into the die. Ideally a three- to four-inch stroke would give operating clearance for loading and handling the components.

It is interesting that most jackets for calibers over .308 are suitable for the caliber if they are about an inch long. This seems to produce a good weight of bullet for all calibers up to the .458 rifle. The .50 machine gun requires a longer jacket although a one-inch jacket flat-base lead-tip bullet could be used for experimental purposes. There is no good way to trim jackets by hand. All the ways are slow and less than satisfactory. Tubing cutters are not very accurate and roll in the jacket edge. Hacksaws and jeweler's saws are tedious and leave somewhat burred edges. The best way to trim off jacket material is with a rotating thin-edged cutting tool in a turret lathe or automatic screw machine. A quick-acting air or lever collet closer permits fast feeding of the tubing in the manner of bar stock. A positive stop on the turret fixes the length of the jacket. Soft copper will require a sharp tool with good relief. A regular cut-off tool may be used but more satisfactory work results from a special thin-edged tool or a rotary cutter.

An engine lathe can be used, either with collets or by carefully closing the three-jaw chuck so it does not crush the tubing. Power feed on the cross-slide helps prevent hogging of the tool and makes long runs more bearable for the operator. The ingenious handloader can probably devise other ways to cut tubing in small quantities, and since larger-caliber rifles for which such tubing is used are not often shot frequently, slower production is probably acceptable. Where longer runs are needed, it is worthwhile to pay a machine shop to cut off a thousand or more lengths of tubing instead of trying to trim them by hand.

While trimming copper tubing may be troublesome without the machinery, such jackets have a great advantage over commercial jackets given the political situation of our times; it would be virtually impossible to prevent handloaders from obtaining copper tubing, just as it would be very hard to eliminate all sources of lead. While commercial bullet jackets can be subject to the rigors of wartime shortages and government regulations or taxes on components, copper tubing is so universally used in appliances, automobiles, industry, plumbing, and other fields that even if all new sources were cut off, the junkyards of the country would furnish years of supplies for handloaders.

Chapter 20

Multijacket Bullets

One of the most interesting aspects of home swaging is the production of bullets which have performance characteristics that are designed for specific situations. One such design is the multijacket bullet, in which a smaller-caliber jacket is slipped inside of a larger-caliber jacket, both having seated cores. The two are swaged to make a single bullet with two jackets. If the inside jacket is inserted with the base up, toward the mouth of the larger jacket, a superior hunting bullet can be made. The tip design can be made for expansion at lower speeds and longer ranges, while the solid base of the inner jacket will remain intact even if a shot is taken at close range and the forward half of the bullet is shattered.

The old German H-Mantle bullets had a fold in the jacket walls to form a sort of wall in a similar manner, but they were far less effective than the double-jacket design. A partitioned bullet is similar, but much more difficult to make at home. Since the lengths of both cores and jackets can be varied to form different ratios, the method provides a wide range of control over the terminal performance at different speeds. A handgun bullet with a partition can be made in a similar manner, something not available any other way. Such bullets can have the open tip required for good expansion and yet still retain a deep-penetrating solid core covering should the tip fail to hold together on a close-range shot.

Multijacket designs can also be made so that a full metal jacket results with no opening in either base or tip. The process simply involves using a pair of short jackets end to end over a core made just the right weight. If the weight is correct, the two jackets will just touch ends when the bullet is finished. Such a bullet is not like a regular full-patch military bullet, in that it can expand, but in the center where the jacket is divided and thinner. This technique makes it possible to swage large-bore rifle bullets using handgun half or three-quarter jackets end to end. Pairs of half-inch-long .38 pistol jackets, for example, make an excellent 200-grain .375 caliber rifle bullet. A 350- to 400-grain .458 rifle bullet can be made using .45 caliber handgun jackets end to end. A .400 magnum or .405 Winchester bullet results from a pair of .41 caliber jackets end to end, though they may have to be drawn down below original size first.

The unusual full-jacket but expanding bullet is low in cost and may represent the easiest and best kind of jacketed bullet to swage, since the tooling required is nothing but regular swage dies. Copper tubing is an alternative jacket source, but it requires some extra time and tools.

Two jackets can be telescoped together to limit expansion by eliminating the weak waist. Jackets which will slip within each other, depending somewhat on the way they are put together and the amount of lead exposed at the tip of the inside jacket (some lead should be exposed on the inside jacket so that there is no air gap between the jacket mouth and the base of the outside jacket) plus their thickness, are as follows:

The .224 jacket will usually slip inside of .257, 6.5mm, and .270 jackets.

The 6mm jacket will fit .270, 7mm, and .308 jackets.

The .257 jacket will fit most .308 and 8mm jackets.

The 6.5mm jacket will fit 8mm jackets.

A .30 short jacket will slip into a .357 or .38 caliber handgun jacket.

An 8mm short jacket will also fit .357 and .38 caliber handgun jackets.

The .357 handgun jacket makes a reasonable fit to .41, .44, and .45 handgun jackets.

Usually, the use of multiple jackets in handgun calibers means

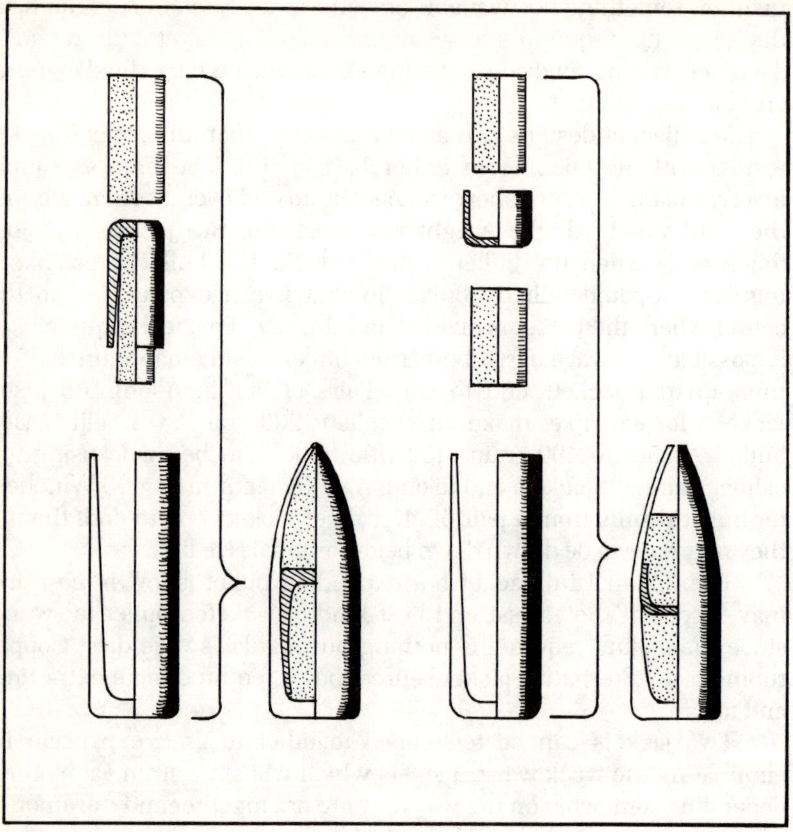

Multijacket bullet construction.

using a half jacket, which is about a quarter inch long, inside a three-quarter jacket or full jacket. These run from about 0.4 inch to 0.55 inch long. The same jacket is normally used for 9mm, .38 Special, .357 magnum, .380 ACP, and similar diameters; so bullets for these calibers can use a short .30 jacket or a .308 gas check as a partition. The core is seated in two steps. First a normal but short lead core is seated inside the jacket. Then the gas check or short jacket is placed with the mouth up, toward the mouth of the larger jacket, and a lead core placed over it. Pressure is applied to the lead core either with a punch that slips inside the outer jacket, for open-tip designs, or with a full die-bore size punch for lead-tip bullets.

Usually, in rifle lengths, the inner jacket is seated with a lead core in a die set for that caliber. That is, if a 6mm jacket is to be used as the inner filling for a .308 partitioned bullet, then a short 6mm jacket is filled with lead in the usual manner using a 6mm core-seating die. The lead is made longer than the jacket by a small amount, and a punch the size of the die bore is used to seat it. Then this 6mm seated core is placed with the lead facing down, toward the bottom of the larger jacket. A short lead core is placed on top of it, and the assembly swaged in the normal manner in a .308 core-seating die. It is possible to do away with the smaller caliber core-seating operation if a very short jacket is used, so that it acts more as a gas check than a complete jacket in itself. If sufficient pressure is used, it may even be possible to make the bullet by dropping all the components in the proper order into the larger jacket and swaging in the normal way. But there is a good chance that the shank and base of the bullet will be undersized or will have an hourglass or "wasp-waisted" shape.

An interesting variation on this design is the use of a fired small rifle primer for a partition in a .224 bullet, or a large rifle primer (fired, of course) for a partition in a 6mm. For most accurate results, the anvil should be pried out and just the primer cup used. The flanges should point toward the jacket mouth. A short core should be seated inside the jacket first, then the primer cup dropped into the jacket, and another short core seated on top of it. The flanges will expand out to touch the walls of the jacket and form a partition between the two cores. For a really high-performance partition, the lower core and primer can be fluxed and soldered into place! This can be done by tinning the inside of the jacket with a low-temperature solder, tinning the primer cup, and after swaging, heating the assembly with a hot air gun or torch until the solder melts. Then the assembly is reswaged to squeeze out any gas bubble that may have formed.

Solid-point bullets can be made by using a short, smaller jacket at the very end of a larger-diameter jacket. A step shoulder results, similar to some of the military .303 British bullets. For really deep-penetrating bullets, more than one internal jacket can be used. As many as two separate partitions can be formed by putting various combinations of jackets together and filling them with short cores. To make a longer-than-normal bullet for a given caliber, without going to jacket redrawing dies, one can either place two same-size rifle jackets end to end over a proper length core as described with handgun jackets in large rifle calibers, or place a smaller jacket inside a larger one with the base of the smaller one forming the nose.

The use of gas checks and short jackets inside of copper tubing, to form a solid base without expensive tooling, is covered in chapter 19. The use of gas checks or short jackets in stacks, to form a multiple-projectile bullet assembly, can be done easily if the nose and base shapes are matched in a pair of punches used with the core-seating die. By making the nose and base so that they will fit easily over each other, a very short section of lead can be swaged into a gas check or short (half-length) pistol jacket. A number of these jackets and cores then can be swaged together as a bullet, with or without a separate outside jacket. If no outer jacket is used, the load usually must be put into a case with filler to support the bottom of the bullet, and with the entire stack protected within the case. If an outer jacket is used, the load will normally hold together when fired until it strikes a fairly solid object, whereupon it normally jars apart and loses much of its penetration ability.

Another variation of the multiple-jacket bullet is the use of a jacket pair that are filled with lead shot instead of a solid lead core. If a lubricating oil is used on the shot, they will not stick together very well even with the highest swaging pressures. A small jacket such as a .44 jacket can be telescoped over a .357 3/4-length jacket filled with number seven shot. More than a jacketful of shot can be enclosed if the .44 jacket is placed inside a short paper roll and shot stacked higher than the edge. The .357 jacket can be wiggled down over the stack of shot until it encloses it completely. This assembly can then be swaged in the normal way using a pair of flat-ended punches. The slug can be swaged again in a point-forming die, making what appears to be a conventional sort of full metal bullet with a little seam around the ogive. The idea of this bullet is to have a slug which breaks up more easily than a solid-core bullet, so that if it strikes concrete or metal there would be less danger of long-ranged fragments flying about.

The use of more than one jacket on a single bullet opens up new design fields for the handloader, limited by imagination alone. Many of the designs are not of any practical use, but a few may well prove superior to any other alternative for a given purpose. It is one of the attributes of swaging that such experiments can be undertaken without costly additional tooling.

Chapter 21

Specialized Bullets For Handguns

Bullet swaging gives the handloader a unique ability, with certain calibers, of exceeding the factory performance levels usually associated with the cartridge or gun. An example is the .25 ACP, which is normally considered to be only as effective as a .22 short for game or defense. A big part of the reason for this larger bore's poor performance in relation to most of the .22 rimfires is the bullet. Of course, the gun has a lot to do with it: any of the rimfire cartridges shot from a two-inch barrel instead of six inches or longer would certainly prove much less effective. But the handloader can make a much better bullet and gain significant improvements in the performance.

The .25 ACP usually is loaded with a full metal jacketed, round-nose bullet rated at 810 feet per second out of the 2-inch barrel. This is 73 foot-pounds of muzzle energy. A .22 short, fired from a 6-inch barrel, is rated at 1,035 fps for the high-velocity loadings, developing 69 foot-pounds of muzzle energy with the 29-grain lead bullet. A .22 long rifle with a 40-grain bullet gets 1125 fps, for a muzzle energy of 112 fps, again from a 6-inch barrel. Although there are no 6-inch barrel .25 ACP pistols except perhaps for custom-made guns, both the velocity and effectiveness of the bullet when it strikes can be improved.

The first step is to decrease the bullet weight slightly but to increase the bearing surface and make certain the bullet will seal the gas in the bore of the gun. With so many makes of lower-priced defense type pocket pistols on the market, bore sizes do not always come out precisely at .250 to .251 inches. The handloader needs to bear in mind that the .25 caliber rifles are usually .257 inches for the bore plus rifling depth, so that jackets or gas checks for the .25 rifles do not work in the smaller .25 pistol components. A bullet of about 40 to 45 grains can be made with a dished or cup base, and with a slightly modified Keith-type nose. This bullet will have enough bearing surface to align itself in the bore, even though it has less weight than the factory bullet.

Factory bullets and the one cast bullet available at this time both have very poor ratios of length to shank, or short bearing surfaces that do not permit the bullet to align itself properly in the bore. The result is a large amount of wobble as the bullet is pushed through the barrel. Since the guns usually have poor or no sights, the accuracy is not considered too important. But even a .25 ACP can be shot more accurately if the bullet fits the bore and is stable both in flight and while it is being accelerated. The cup base that permits the lighter bullet to have a longer shank also lets the gas expand the base section and help seal the bore, so that all the gas pressure is used to propel the bullet, and none of it escapes to weaken the force of the cartridge. Two improvements are gained by this one design feature.

The nose of the bullet has a great effect on how quickly it will give its energy to the target. In defense and small game shooting, it is desirable to have the most energy given up in the shortest time, so that the force will change the shape of the bullet from a low-resistance ballistic form to a very high-resistance shape. This is sometimes called the "shock" effect of the bullet, and with small calibers like the .25 ACP there is little enough of it with a round-nose bullet shape. Although speeds of from 900 to 1,000 feet per second can be reached safely by handloading the 40- to 45-grain bullet, these are still not great enough to expand a fully jacketed bullet. Even with a hole in the nose, the jacket will resist expanding.

When the jacket keeps the bullet from expanding, it retains the effective shape that lets it slip easily through the air. This shape also lets it slip easily through the target medium with as little damage as possible, and that is exactly what is not wanted. In larger game hunted with heavier calibers, deep penetration is a must, but for defense with such short second-choice type weapons penetration is secondary to immediate damage and a larger wound capable of incapacitating a criminal as fast as possible with the caliber being used. A soft lead tip will expand more easily, but even pure lead bullets don't always expand as well at speeds below 1,000 feet per second. The shock effect, if there is to be much of one, will have to come from the initial shape of the bullet nose.

The single straight-walled swage die used with a Keith-type nose punch will produce a bullet with a sharp step between the shank and start of the nose. This step may be from .020 to .045 inch wide. In a long series of tests using all the usual styles of bullet noses and many experimental types, conducted by the Corbin brothers in 1972, it was found that the small step around the nose had more effect on wound size and amount of secondary projectile movement from the target material than any of the various cavities or trick nose shapes, since most of these did not change or induce expansion to any significant degree at speeds below 1,000 feet per second. The few shapes that did show significant improvement did not work in most autoloading pistols.

A .25 ACP could be made to create the same size of wound cavity as a .32 ACP factory load or any of the other revolver or pistol loads in the same general caliber and velocity range. This did not mean it imparted as much energy or did as much total damage, but the wound cavity could at least be the same general size. This is a significant increase in performance over factory bullets. By the same token, a .32 ACP could be made to nearly duplicate a .38 Special in wound cavity diameter and depth, using the same bullet design. Again, the two calibers do not impart the same energy, but the effect in regard to shock can be made almost the same if one uses a round-nose bullet while the other uses a bullet designed to get the very most out of the gun and cartridge safely without excessive powder charges or pressures.

Bullets for the .25 ACP do not have to be jacketed. They can be swaged of lead and shot unlubricated. Some fouling will occur, but at the low pressure and speeds, and with the short barrel, it isn't hard to remove. A .243 rifle gas check can be swaged to the base of the bullet simply by dropping the core into the gas check and swaging the two normally. The gas check will expand to .250-.251 inch from the swaging pressure. A jacket can be made for the .25 ACP using .22

short cases, or an even better jacket can be made using fired shotgun primer battery cups.

The shotgun shell uses a primer with a small cup and no anvil set into a larger battery cup, having the separate anvil pressed into the bottom of it. A flange on the battery cup keeps the assembly from being pushed too far into the shotgun shell head. A set of tools was developed by Richard Corbin in early 1977 to take the shotgun primer apart, draw the battery cup into a straight-walled jacket for the .250–.251-inch bore, and then seat the core into this cup so that it becomes a completed Keith-type bullet of very great effectiveness. For guns that will not feed the Keith-type nose, a second die will turn it into a standard truncated conical or open-point expanding bullet with lead tip, but having the jacket brought partly up into the ogive. Such bullets do not lead the bore.

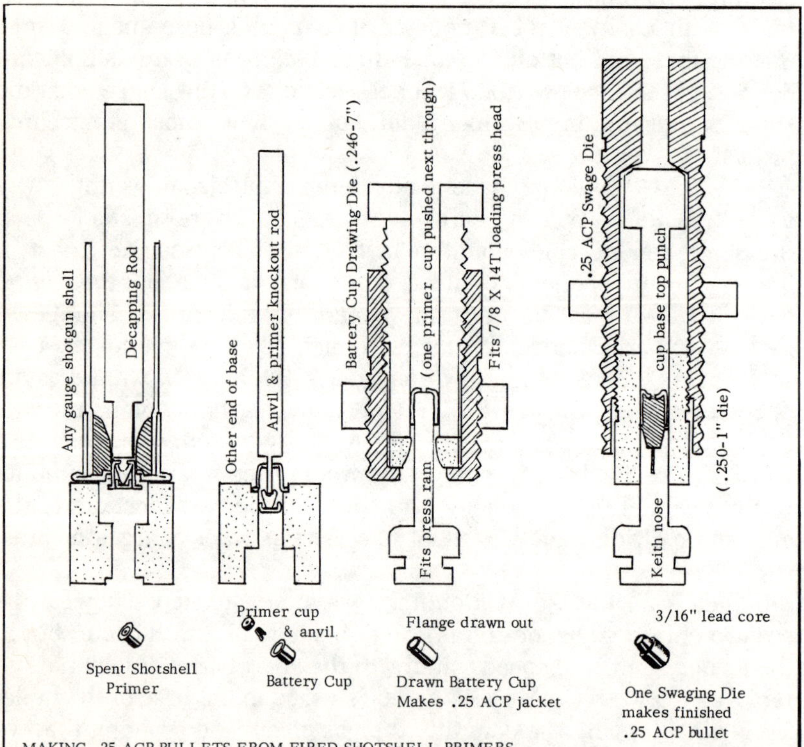

MAKING .25 ACP BULLETS FROM FIRED SHOTSHELL PRIMERS

Making .25 ACP bullets from fired shotshell primers.

The tools are simply a depriming rod and a base to hold the shotgun shell, a second series of holes on the other side of the base that holds the battery cup and lets another punch be pushed through the flash hole to remove the primer and anvil, and a draw die with a punch that presses the battery cup through it to remove the flanges and size the jacket. If desired, the spent primer can then be put back into the battery cup, upside down, so it seals the flash hole. This isn't necessary but does make a neater-looking bullet and prevents any lead from being melted by powder gas (not likely with the low pressure and short barrel).

At this writing there are no other jackets available for the .25 ACP. The .22 short can be drawn out to make a .224 caliber jacket, but with the mouth flared to .250 inch so it will seal the bore of the swage die. A lead core is then inserted, and when pressure is applied, the lead will expand the rest of the jacket to meet the die walls. If the mouth were not expanded or belled first, it would be possible for the mouth to expand later than the rest of the jacket, so that lead might already be extruded past the punch and start seeping between the jacket and the die. If the core is seated below the jacket mouth, it is not possible to expand the jacket mouth during core seating in any case, and the mouth would have to be belled first.

Usually it is necessary to make lead-tip bullets only, using the .22 short in this manner, in order that the pressure can be evenly applied all along the jacket walls by having a long enough lead core to contact the entire jacket length. This means lead will protrude beyond the jacket end, and a full bore-size punch will be necessary to seat the core. A second die which forms the ogive (point-forming die) is used to form the lead-tip bullet with the jacket brought into the ogive. The shotgun battery cup is the ideal diameter after passing through the flange removal draw die, and any style of bullet can be made just as with a commercial jacket.

Another caliber that needs help from the handloader who swages bullets is the .44 Special. This fine old caliber is underloaded and the bullets that have been used with it almost universally are of the styles that don't let it develop full potential for defense use. If a bullet is needed that will expand rapidly and not penetrate deeply, so that innocent people are in less danger of bullets passing through walls or bouncing off sidewalks, yet a large caliber is required for the utmost safety of the person doing the shooting, then a .44 Special is hard to beat. The small frame, thin-walled cylinder pocket .44 pistol will not stand up to hot handloads in any case, and a moderate velocity is necessary for these guns. But the bullet can be designed for expansion at moderate speed.

One such design is the three-quarter-jacketed cup point, with a deep hollow cup that both makes the bullet lighter for its overall length and thus lets velocity be higher without extreme recoil or pressure, and also makes it possible for lower speeds to expand the bullet to an extreme degree. Such a bullet is made by replacing the usual cavity-type nose punch in the swage set with a machined projection that pushes deeply into the core as it seats the core in the jacket. A cavity is formed that is the shape of this projection on the punch. The resulting bullet looks like a wadcutter from the side. If the lead core is to be completely within the jacket, the punch needs to have a section that will slip snugly into the jacket mouth. If the core can be just a little longer than the jacket, the punch can be full diameter for the die bore.

Either of these bullet styles (lead-tip cup point or full-jacket cup point) can be further formed into a more conventional style of ogive by partial swaging in the point-forming die. The jacket is brought around smoothly to a wide-open tip, while the cavity in the bullet remains essentially the same diameter at the bottom and closes slightly at the top. This creates a large pear-shaped cavity in the bullet. In truth the full cylinder shape will expand more quickly at lower speeds, but the looks and ballistics of the bullet formed in two dies is superior. Much of the appeal is in the appearance, but performance with any of these bullets is much better for home defense than the hunting type bullet or round-nose, designed for higher speeds or deeper penetration.

Seating a cup-point bullet can be a problem, especially the lead-tip full-cylinder type. It expands so easily that it can start expanding during the seating operation. To prevent this, the bullet-seating die of the reloading set should be equippped with a custom-made punch that duplicates the core-seating punch shape. A fairly snug fit would be best between the case mouth and the jacket, rather than a heavy crimp. Special taper crimp dies that put a gentle taper in the mouth of the case and compress it against the bullet after the bullet is seated can be used. These are very much like a jacket-drawing die, consisting of a ring with a precise diameter of hole, held in a pair of threaded parts that fit the press head and clamp the ring die firmly between them at the top and bottom. The cartridge is simply run into the mouth of this die so that some slight compression is applied to the jacket and bullet, and at some point along the tapered area of pressure application the bullet will spring back more than the cartridge case, resulting in a ring around the mouth which, although it cannot be seen or felt, is slightly tighter than the fit would otherwise be. The same die can be used to roll the mouth into a cannelure very gently, or to help iron out excessive belling of the case mouth.

Another example of a special bullet that the swager can make to solve a performance problem is the .22 caliber black powder pistol or rifle bullet. Several guns have been introduced that are made to shoot an airgun pellet or a .22 caliber round ball using black powder. There is even a pepperbox revolver available at this time. Airgun pellets are not really the ideal projectile for this, and usually they are expanded so much at the base by the muzzle blast from any reasonable load that they cannot fly accurately. A swage die using only the core-seating die and a pair of punches can turn out superior muzzle-loading bullets. One punch has a projection to make a deep cup in the base. The other has a Keith-type cavity to make a very fine target or small game nose shape. This bullet is similar, but with a much deeper cavity, to the .25ACP or .32 ACP bullets discussed. The edges of the cavity in the base are made thick enough to resist the deforming effect of the muzzle blast.

An interesting way to handle this job is to use the regular core-seating die from a .224 rifle swage set. The top and bottom punch are changed for the two punches mentioned. The bore of the core seater is usually about .2235–.2238 inch in diameter. This makes a bullet that fits and loads easily in the .22 muzzle loaders, yet expands just enough to seal the bore and give highest possible speeds for the powder charge. The shooter with a .224 rifle swage die set needs only add a pair of punches and use part of the set for making these bullets. Another use for them is in air rifles. The accuracy and speed of very light .2235–.2238-inch bullets of this design has been tested and found excellent in most cases. It is usually superior to the tins of pellets one purchases for ordinary shooting. Since the weight is up to the person making the pellet, and since weight is about the only factor controlling the velocity with a given air pressure, flatter-shooting airguns can be a result of the experiments.

Another experiment with these bullets, which use no jacket of any sort, was conducted using a solid copper cartridge turned on a lathe to duplicate the .222 Remington magnum, but having only a neck and a flash hole and primer cup connected by a small-diameter port. The case was solid, in other words, with no powder room. All the energy of the primer was directed to the neck of the case, where a lead bullet of the type described could be pushed by finger pressure into the cartridge. It would seat against the bottom of the hole at the same location each time.

The primer pocket was made large enough so that the primers could also be seated by hand and pressed out easily with a small-diameter rod. Reloading this cartridge took only seconds and it could be fired over and over again quite rapidly. The cartridge was used in a fine benchrest rifle, complete with a good target scope and heavy match barrel. The purpose of the combination was to attempt to find out if a target shooter could reasonably expect to practice indoors, during the bad weather and off season, and maintain his edge of control by using a reduced target at closer range in the comfort of his home.

The combination worked out quite well. Drop was a little more than might be expected from a 100-yard projectile impact at normal velocity; so the sights would have to be reset. But the accuracy was good enough so that a shooter could actually see if he was maintaining his skill. In other words, trigger control, breathing, grip, position, and other factors of shooting could be kept honed to a good edge all winter long without leaving the garage or even living room. A backstop consisting of a cardboard box filled with burlap and placed against a wooden plank would be sufficient. Any regular .22 indoor bullet trap would also be more than adequate, since the nose and velocity were less by far than even a .22 short.

Pellets seated backward in the case did not work nearly as well, although they could be used to shoot pests in the backyard. The pressure required to swage these small-size, all-lead bullets is so low that a hand press could be used—one not mounted to the bench, but operated by moving two levers held in the hands. A press was designed and tested, and it functioned well enough to make jacketed .224 and 6mm bullets without undue strain. The press was essentially a tool like the power ejector that fits over the top of a swaging die and uses compound leverage to press the ejection rod down. A half-circle milled out of one side of the body allowed the bullet to be removed and components to be inserted into a die screwed into one end of the tool, while the punch and its floating holder were stationary at the other end. The die moved with the ram, so that ejection was automatic on the back stroke.

This press as well as the solid cartridges were merely experimental work done in 1975 by Richard Corbin, and did not result in tools being produced, but they do show some of the applications that can be developed from special bullets. Handgun bullets offer a great deal of

Fired factory autoloading pistol bullets show almost no expansion in any of the tested target mediums.

Autoloader bullets made at home, fired into same targets as previous illustration.

fertile ground for experiments since handgun ballistics have been ignored in comparison to rifles. Not until the original Super-Vel bullet was a concerted attempt made to improve the most critical factor in handgun performance, the characteristics of the bullet itself. The handloader has the power to continue these improvements through innovative designs of his own, through the vast capabilities of bullet swaging.

Chapter 22

The Hoxie Bullet

The May 11, 1907, *Scientific American* contained an article which is of interest to bullet swagers, because it concerns a very simple technique for making an expanding bullet that appears to be a solid-point round-nose projectile. While some of the claims are extravagant, there is enough basis in the design to warrant experimentation. The following is a direct quote from the article:

> The small-bore, high power rifle that is in general use today was originally designed as a military rifle, the object of which is to wound or maim at extreme range; and owing to its high velocity, the trajectory is so flat that the raising or changing of sights under ordinary circumstances is unnecessary. Sportsmen, seeing the advantage gained by great velocity, were quick to adopt this type of rifle. It was found that by inverting the jacket of the full metal-patched bullet so as to leave the soft nose exposed, this bullet when striking hard substances such as bone, will very often mushroom or expand, causing a severe wound. Improvements in powder have from time to time increased the velocity of these bullets until they now have a muzzle velocity of 2,700 feet per second. With this velocity even the

> soft nosed bullet will pass through the animal without expanding in the least until some hard substance is struck, when it is apt to fly to pieces. The great heat caused by friction in the air causes the bullet to cauterize the veins and arteries, causing little bleeding and thus making it impossible for the hunter to track his game by the blood. It is stated on reliable authority that this year in Nova Scotia over forty per cent of the game hit or wounded escaped, some to die a lingering death in great agony.
>
> To overcome these objections to the ordinary bullet, Mr. G. H. Hoxie, Chicago, Ill., has invented the form of bullet illustrated in section in the accompanying drawing. The bullet consists of a jacket with a filling of lead in which a steel ball is seated. In another construction a steel wedge is used in place of the ball. Behind the ball is a chamber formed in the filling. When the bullet strikes an object the ball is forced into the chamber, expanding it and tearing it open. One of the illustrations shows the character of wounds produced by this bullet as compared with other wounds. The arrows 1, 2, and 3 point to wounds produced by the ordinary soft-nosed bullet. The wounds made by the improved bullet are four or five times larger, and need no designating arrows.

Although the Hoxie bullet did not prove up to the very amazing claims Mr. Hoxie made for it, there is still a range of velocity where a ball seated in a cavity will actually help expand the bullet. At slower speeds, the mechanism fails to expand and at higher speeds the bullet goes to pieces just like any other of the same jacket construction. The main advantage to a handloader who is swaging bullets is the protection of the very tip of the bullet in the magazine, and the finished appearance of the tip achieved without great difficulty. A hard lead shot can be substituted for a steel BB shot if desired. When the tip is formed, this steel or hard lead shot will be swaged into a tip less likely to deform than the pure lead core. In automatic pistols and magazine rifles this is an advantage which could well be considered.

To quote from the Hoxie advertising of the day is interesting, especially if viewed from the more mature experience of today's shooter and handloader. Bear in mind that this was 1907 and few hunters had much conception of terminal ballistics other than their experience with cast bullets or factory-jacketed bullet loads, or full-patch military ammunition. Quoting from the Hoxie advertising booklet:

> In this bullet, a great improvement has been made over any form of the mushroom type bullet heretofore used. It will upset to three or four times its original size upon the slightest contact with animal tissue. The hide of a deer will cause it to tear a wound where it enters in most cases larger than where the ordinary soft nose bullet passes out. Hunters of large game and guides will at once appreciate these

> results, as it makes the smaller calibre (viz.: 25-36, 30-30, 33, etc., so popular for their accuracy) powerful enough for the largest game.
>
> The metal jacket is so thoroughly expanded that lung or stomach shots are sufficiently effective to drop an animal in its tracks or nearby. Thus, the hunter using the Hoxie Bullet will be sure of his game without a long chase.
>
> In "Outdoor Life," F. W. Chase, Cody, Wyoming, cites an instance of killing an old grizzly "four toes." The bear was wounded eleven times by the guide and hunter, the shots taking effect in the mouth, shoulders, legs and liver; the last shot, a lucky one, which landed in the mouth, stopped the animal. Had a Hoxie Bullet been used, any one of the shots would have given the desired result. This is one of many instances of like nature that are being brought to notice from time to time, and apart from torture to the animal, damage to the hide and meat, economy must be considered, for, while this bullet may make the ammunition cost more, a considerable savings in the end is very apparent, and there is much less liability of running out of ammunition.
>
> The inventor of this new bullet has had a wide experience in hunting large game, has made exhaustive tests and extensive experiments, and has confidence that when his bullet is once used the experienced hunter will recognize its superior merits.
>
> The bullet is of equal weight of standard bullets and has been tested by rifle experts at target, and the accuracy is in no way impaired.
>
> The Hoxie bullet will appeal to all sportsmen who are in quest of large game. It will double the killing power of your rifle. The inventor has made exhaustive experiments, and has discovered a way by which a bullet of 30 caliber in diameter will have sufficient stopping power for any game. It does not impair the velocity or accuracy, and will mushroom on contact with the slightest resistance, the hide of a deer alone being sufficient to expand it three or four times its original size. Our bullet can be depended upon to expand and kill no matter how soft the tissue hit. For instance a paunch or neck shot (below the spinal column) has often resulted in instant death when the Hoxie bullet was used. Game is getting so scarce, or driven so far back as to be accessible only after long journeys and considerable expense. Hence the up-to-date hunter wants to be equipped with everything that will give weight to his position. Every large game hunter has had the experience of having the "big one get away, badly hit." Such an experience prompted the inventor to try and find a way whereby our new high-power rifles could be made to be depended upon, and after three years of careful study and experiment he has developed this bullet to that state of perfection where we can unequivocally recommend it to our fellow sportsmen. After you have tried them and been convinced, please let us hear from you.

The advertising contains the usual signed testimonials from people who have used the bullets, with glowing reports on the success. This was typical of the advertising of the day and was probably honest enough, since even the worst bullet is bound to produce a very impressive kill at least once in a while. The advertising went on to interesting extremes, however, which are given here because they are so humorous in light of the information we have today on bullet performance:

> A Word to Beginners. Give Hoxie Cartridges a Fair Test. Do not expect simply by getting an accurate rifle and loading it with the Hoxie Mushroom Bullets that all you have to do is to go out into the woods and fire a few random shots and then hire the nearest farm wagon to haul in your game, as it is absolutely necessary to hit the game with the Hoxie bullet, and it will do the rest and do it quick. "One Hoxie shot does the work."
>
> At present we are applying the Hoxie patent bullet to all regular makes of ammunition, loaded with soft-nosed metal-jacketed bullets. We do not disturb the cartridge in the least in applying the invention. The bullets are not removed. Consequently, the cartridges leave our hands in the same condition as when we receive them.
>
> The Hoxie bullet can be successfully applied to any soft nose or even full lead bullet, giving results three or four times as satisfactory as can be obtained from regular factory bullets. We illustrate only a few of principal sizes, but order any style or make you prefer and can be promptly supplied. All Hoxie cartridges have appearance of regular cartridges, but Hoxie impact ball can be seen at extreme point of bullet.

There are two techniques the home swager can use to experiment with Hoxie style bullets. First, one can seat the core using a punch with a conical or hollow-point projection. The ball, which can be a lead shot or a steel BB in the larger calibers, is simply dropped into the cavity and the bullet point is formed in the point-forming die. A little care is needed to keep from forcing the hard ball too far into the point of the die cavity. A second technique is to seat a core just a bit shorter than usual for the jacket length, and then only partly form up the ogive. The bullet is ejected while there is a considerable size hole left in the tip, and the ball dropped into the open area at the tip. Then the bullet is swaged some more, until the ball is brought into line with the jacket mouth but so its largest diameter is below the mouth.

Either technique requires some experimenting with core length to position the ball so it won't simply fall out after swaging, but once the correct combination is found a number of Hoxie type bullets can be made in a hurry. Regardless of whether such a bullet makes small calibers adequate for the largest game (certainly it will not!), the appearance is pleasing and the tip protection is worth considering. The round ball finishes the tip of the bullet without the necessity of using a tip-forming die after the ogive-walled point-forming die operation.

Chapter 23

Uses and Limits Of Draw Dies

While swaging is the process of expanding the diameter of a lead core and jacket by pressure on one or both ends, drawing is the process of reducing the jacket, core, or bullet diameter by pressure against one end and the sides. A drawing die is open at both ends, and consists of a narrow ring die with a constriction that is the correct size for finishing the bullet or jacket as desired, and a guide cone which is a carefully worked out approach to the constriction. The ring die is held in an adapter for the kind of press in which it is to be used, and a matched punch is used that fits both the jacket interior and the constriction diameter in the ring die—or else, for bullets or cores, the punch simply slips through the constriction.

Draw dies are used most frequently for either drawing a jacket down to the next smaller size, and lengthening it in the process, or for reducing the diameter of a finished bullet slightly, to take out any taper in the shank and to control the final diameter very precisely. Sometimes a draw die is used to make a finished commercial bullet into one of smaller caliber. This can be done within certain limits. If too big a reduction is used, the bullet shape will be distorted and the bullet may

A jacket draw die.

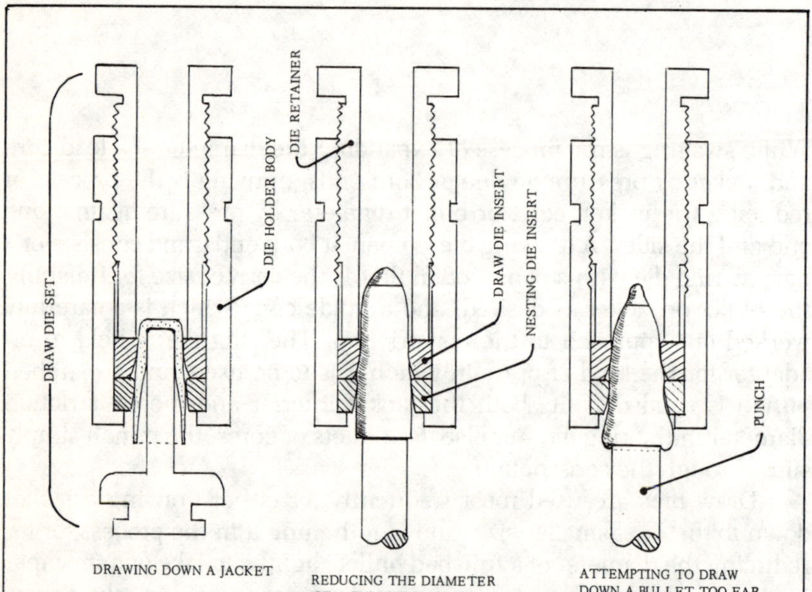

The operation of jacket and bullet draw dies.

not even go through the die. Quite often a shooter will want to order a simple ring die and punch so that he can reduce a commercial bullet to the next smaller caliber, especially if it is an odd size that can't be purchased. Each case needs examining and test bullets need to be received before final word is given, but here is what can happen to spoil that idea:

First, if the reduction is great enough, the force necessary to reduce the bullet will be greater than the force the jacket can stand. Instead of pressing through the die and coming out the top a smaller diameter, the bullet base merely mushrooms. To eliminate this, one can make several dies that reduce the diameter only a few thousandths of an inch per draw. This would seem reasonable, although the cost becomes higher and it may be better just to order swage dies, jackets, and a press to make the bullets from scratch. However, even this technique has its limits.

There is more going on within the bullet than just shrinking the diameter, whenever it is put through a draw die. The lead within the bullet does not disappear when the diameter is reduced. It has to go somewhere. The base is closed; so the only place the lead can go is out the nose. All the lead that was in an imaginary cylinder, representing the old diameter less the new diameter, is moved forward into the ogive. This can do three things to the bullet. It can increase the bearing surface, which is done because the lead moves the ogive out toward shank diameter. Also, a great enough reduction catches the ogive against the ring die constriction and starts the shank higher on the bullet, closer to the tip. Even if internal lead pressure did not expand the ogive and bring it closer to shank diameter, the reduction itself will result in more bearing surface. The effect is slight for small reductions, meaning only a few thousandths of an inch. Above ten- to fifteen-thousandths reduction the effect starts to become excessive for many bullets.

The next thing the displaced lead can do is to change the ogive shape, making it more round-nosed and less of a spire, spitzer, or whatever shape was originally selected. In some cases the lead pressure changes the nose shape completely. If the shooter wanted to keep everything the same and just have a smaller bullet diameter, he may be in for a sad surprise; and there is no way that the diemaker can guarantee anything about the bullet he will wind up with, except that it will be the diameter the die is supposed to make. With swaging, such problems are eliminated since the shape can be whatever is desired along with the diameter. Still, as long as a person bears this in mind, drawing may be practical for certain bullet shapes.

The last thing that drawing down the bullet can do, as a result of excessive reduction and displaced lead, is to spurt lead right out the bullet nose. The problem with this is that it doesn't always spurt evenly or the same amount; sometimes the lead will expand the jacket at the ogive and other times it may extrude from the bullet tip. When this kind of pressure is inside the jacket, it is usually very close to the point where the base mushrooms or the ogive becomes swollen larger than the shank. Such bullets aren't usable. They are ruined and must be thrown away. This is not to say that drawing bullets is always undesirable. In some cases it is very practical. For instance, a person wanting to make a 9mm bullet of 0.355 or 0.356 diameter out of a .38 bullet of .357 diameter can simply purchase a draw die. The result will be very acceptable, and much cheaper than buying a whole set of 9mm swage dies if one already has the .357 die set.

Other examples include making a .223 Hornet bullet out of a .224 diameter modern centerfire .22 bullet, or taking all the taper and pressure ring off any caliber of swaged bullet, for use in autoloaders or machine guns. Any gun that uses fairly rough feeding pressure against the bullets should be fed with ammo having parallel shank bullets, not match-type swaged bullets with the pressure ring. The parallel shank does tend to increase pressure slightly, but it also eliminates the setback of the bullet into the case from feeding. Even in bolt-action rifles chambered for short-neck cartridges like the .300 Savage, a draw die used on a swaged .308 bullet to ensure that the bullet is actually 0.3080 all the way from the end of the ogive to the base will help keep the bullet in place during feeding. The pressure ring will expand the case neck slightly as it passes through, and then the case is a loose fit on the bullet shank, which is tapered slightly in most cases. The taper lets the bullet slip back into the case too easily for a short-neck cartridge or an autoloader. Redrawing the finished bullet to a uniform straight shank only reduces the diameter in the largest points, and then only a few ten thousandths of an inch.

If a person has several rifles of the same caliber, but they have slightly different bores—such as a battery of .308 rifles that seem to shoot best with bullets that range from 0.309 to .3075 diameter—it is wise to order a swage die set in the largest caliber size (in this case, a finished bullet of 0.309) and a draw die for each different smaller rifle bore. This does not mean that a person can shoot drawn .308 bullets in a .270 rifle, or any large reduction similar to this, but a few thousandths of an inch just to match the bullets to the rifles is quite good technique.

Of course, the most publicized use for draw dies in recent times has been the introduction of draw dies to make .224 or 6mm bullet jackets out of rimfire cases (.22 long rifle). This is nothing more than a special use for the standard jacket draw die. The head of the case has to be enclosed in the guide cone area before any pressure is applied, and the design of the angles and polish of the die has to be held to very good tolerances. The punch that fits the jacket is very critical in its diameter. If all these things are taken care of in manufacturing, the die sets will produce very fine .224 jackets in approximately 0.705-inch length, which is the same as the popular commercial jackets for 53-grain match bullets.

A little less known but still interesting is the use of a similar draw die set to turn fired shotshell primers (the battery cup) into fine .25 ACP jackets. A shotshell primer has a regular design of primer cup, with a large separate anvil, both pressed into a fairly large cup having a flange. If the fired primer cup and anvil are pressed out of the battery cup, using special tools made just to do this job, the little top hat of a battery cup can be drawn through a ring die exactly like a rimfire case. The flange is made part of the jacket wall, and the battery cup is turned into a very fine .246-inch diameter jacket for the .250-inch bore, .25 ACP bullet. Note that the .25 ACP is a true quarter-inch bore, of from 0.250 to .0251 inch diameter, and the .25 rifles such as the .250 Savage or .25-06 are .257 inch caliber. The same bullets do not work in either caliber of gun, of course. Seven-thousandths is way out of line for any interchangeability.

Draw dies for bullet jackets can reduce the diameter much more than a few thousandths in one pass. In fact, if a little bit of uneven mouth or wall thickness can be tolerated, a die can be produced that will turn a .38 caliber pistol half or three-quarter jacket into a nice three-quarter or short full jacket for a .308 rifle bullet! In recent times, meaning the late 1970s, the sources of short carbine or pistol .30 caliber jackets have nearly dried up. Just about the only good way to make these jackets is to draw down commonly available .38 jackets. The dies can be made so that a very even mouth results only if the length of the jacket is not changed a great deal in the reduction process. But even with some mouth unevenness, the jackets thus made are superior to the usual short .30 commercially available in two respects: they have thinner walls near the mouth and will expand much better at low pistol velocities or carbine speeds, and they cost less.

Another good use for a draw die is to turn a commonly available and low cost .224 jacket into a hard-to-find .20 or .17 caliber jacket. Usually a 0.6-inch-long, or shorter, .224 jacket is used for this. One pass through the draw die will do the trick nicely with very even walls and mouth. Another good use is to make a heavy or long .224 jacket out of a 6mm jacket. Whenever this is done, a person should expect that he will need to find or produce a smaller lead core than he uses for regular jackets. The interior of the jacket will be smaller diameter than normal. Instead of using 3/16-inch lead wire or cores for the drawn .224 jacket, he will have to go to a 5/32-inch wire or core size. The same core swage die and core-seating punch will probably not work for both standard and drawn 6mm jackets in his .224. The purpose of these jackets is to make long-range heavyweight .224 bullets such as those made famous by Clark and other fast-twist rifle experimenters.

Still another good use for a jacket-drawing die is to turn a surplus jacket that is available at reasonable prices into a more usable caliber (smaller) that would cost more. A good example might be an 8mm jacket, which is not very common among swaging enthusiasts and usually is available at low cost due to surplus stocks, drawn down to make a .308 jacket. In any of these operations, it is important to bear in mind that the draw die will have a hole that is a bit smaller than the finished jacket size desired; this is due to jacket springback. This is hard to calculate, based on the differences in jacket materials and anneals; so samples are needed to adjust the die for proper operation. Also, one must remember that the die must make jackets smaller than the bore of the core-seating die, not fully sized for the finished bullet diameter. Thus, a good diameter for .308 jackets would be 0.306 to 0.307 inch.

This is pointed out because some shooters want to use their jacket-drawing die to size bullets, too. This can be done, but the die can't be made so it produces both a proper size bullet and a proper sized jacket. Sometimes the same die set can be used with a different punch for different jackets. As an example, a die set to draw .38 jackets into short carbine .308 bullet jackets can also be used, with a longer, smaller punch, to draw a factory .308 jacket down for home swaging operations. Because most factory operations use a draw die as the final diameter and taper control, they start with same-size or oversized jackets. Buying up surplus factory jackets for home swaging, instead of jackets made for use by swagers at home, can mean that the jackets are too large and will give ejection problems, wrinkled ogives and shanks, lots of stuck bullets, and possibly even result in damage to the punches from excessive ejection or swaging pressure.

A set of dies for swaging is matched so that the core-seat die is just a few ten thousandths smaller than the point-forming die, and if a lead-tip die is used, it is a few ten thousandths larger than the point-forming die. The difference can't be seen very easily, but it can surely be felt in the operation, if it isn't there. The same holds true of jackets. The jackets must be smaller by a few ten thousandths at least than the core-seating die bore. They can be several thousandths smaller, such as a 0.353-inch jacket in a 0.357-inch die, and no harm will result to the quality or accuracy of the bullet. In fact, it may be improved from the snug grip of the jacket to the core. Too much difference, and lead will bleed around the core-seating punch, lower the internal die pressure, and result again in loose cores or an hourglass dimension to the shank of the bullet. Usually, from one- to four-thousandths smaller than the final bullet diameter desired is a good range for the drawn jacket.

When ordering a draw die for reducing a commercial factory jacket to use in a home swaging operation, sample jackets are a virtual necessity. This is because the springback of the jacket, rather than the diameter desired, will determine how large to make the die hole. Also, the internal shape and diameter of jackets varies a great deal from one manufacturer to another, or even within lots from one manufacturer. The punch must match the inside of the jacket properly; at least, it must fit inside the jacket bottom properly. The amount of pressure the drawing operation will take is controlled in part by the fit of the punch to the jacket interior. It can be made too great for a hand press or so easy that a light hand press will do the job, just by this one factor. Of course, the resulting jacket length and wall thickness is also changed by the punch diameter, and this is why the pressure can change so much. The swager must send samples of his jacket, or specify a jacket the diemaker has on hand, to get a draw die that will work properly.

A final note on drawing has to do with the recent surge of interest in metallic silhouette handgun matches. Those matches consist of a series of metal targets in the shape of animals, set up at various fixed ranges, which must be knocked over to score a point. Obviously, the spectator interest can be high and the game is a good deal of fast-paced fun for everyone, but the bullets need to be made in a special way so that they do the opposite of most game bullets. Most game bullets are made to expand so that they present a large wound cavity. Most target bullets are made only with accurate grouping in mind, and modest speeds. The metallic silhouette game requires that every foot-pound of energy possible be transferred to the metal target, and as little as possible be wasted in changing the bullet shape.

The match also requires highest possible speeds consistent with good accuracy, and a reasonable flat trajectory for the caliber. It is one of the few matches, if not the only one, where remaining energy of the bullet can determine a point almost as much as the accuracy of the bullet. Those targets must fall over, or they are not counted as a point. The idea of the game is that it is a practical course of fire, having a good deal more to do with actual game hunting requirements than any paper shooting match. If the target won't fall over, this is indicative of what might happen to actual game; too weak a load, too light a bullet, as well as poor placement, can lose the point.

What has this to do with draw dies? Several hundred shooters have found that they can use ordinary .45 ACP military hard-ball bullets and pass them through a 0.429-inch draw die to make a reasonably good .44 magnum bullet. This bullet does not expand, at least not much, and transfers nearly all the remaining energy to the target for more sure knock-downs. Also, another die can be added to bring the .429 bullet down to a .41 magnum size (0.410 inch) for use in that gun. Recently, new tests have shown that placing a .38 three-quarter jacket over a core, and putting this into a .44 jacket with the bases out on both jackets, produces an even better bullet for the match. A special single die can finish this bullet, and it is only slightly more expensive than the draw die, but the bullets can be made of jackets and lead instead of expensive finished bullets of another caliber. The weight can be controlled also, which can't be done by drawing down a finished bullet. The double-jacket bullet is even better than the open-base .45 bullet drawn to .44 caliber as far as retaining its shape and transferring energy.

Chapter 24

Accessories For Swaging

The basic tools for swaging are: (1) a press, (2) a set of dies, or single die for straight-wall type (semiwadcutter) bullets, (3) a good lubricant, such as Draw Die Lube or lanolin, (4) a core cutter, if lead wire is to be used, or (5) a core mold, if lead is to be cast for cores. The press can be a regular reloading type for small rifle calibers or handgun calibers, a specially modified press with a solid, threaded ram, or a special swaging press. The dies can be as simple as the single die for handgun bullets, or a three-die set for fine rifle bullets.

There are relatively few accessory items needed to make more accurate or versatile bullets. This is one of the economical factors about swaging. The bullet that comes from the final die is ready to shoot, in most cases. It needs no lubrication or sizing. But in some uses, a bullet can use a cannelure. The cannelure (pronounced like "can-a-loor," as you might say "can of worms") is an indentation rolled around the bullet. It can do several things for us.

First, it can serve as a place to crimp the end of the cartridge case, to hold the bullet against spring pressure in a tubular magazine rifle, against feeding pressure in a semiautomatic pistol, or against recoil energy in any gun. The cannelure can likewise be used on the case, behind the bullet, in guns like the .45 ACP pistol which use the end of the case to headspace. This keeps the force of feeding into the chamber from forcing the bullet deeper into the case, and improves accuracy.

But the cannelure can also play a part in controlling the performance of the fired bullet. The cannelure will press a portion of the jacket into the lead, and depending on the depth and shape of the cannelure groove, it can either weaken or strengthen the jacket at this point. In any case, it will act as a point of greatly modified resistance to jacket foldback. It will either let the jacket break off at the cannelure edge or it will stop further splitting of the jacket on impact. Of course, it won't have any noticeable effect at higher velocities than the point at which the jacket is torn past the cannelured area, or on lower velocities than the point at which the jacket foldback just reaches the cannelure.

But in a reasonably narrow range of speeds at impact, the cannelure can either improve or decrease the effectiveness of the bullet's expansion. Which it does depends on whether the jacket is weakened at the cannelure to the point of breaking, or whether it is thickened and strengthened so that the mushrooming of the point is stopped and retained on the bullet. Experience shows that a relatively light cannelure done with a roller having a rounded edge will produce a strong belt of metal as thick or thicker than the rest of the jacket right at the bottom of the groove. A sharp roller with very straight edges will weaken the jacket at the very junction of the cannelure wall and bottom, sometimes to half the strength or less of the rest of the jacket.

The device which rolls the cannelure is a cannelure tool, and such devices are relatively simple to use. A hardened tool steel roller with small grooves in it is mounted on a shaft, all of which can be rotated by a crank. This assembly is placed on a hinged arm of one type or another, so that it can be pressed against a bullet as the bullet rests on a sort of V-way made up of two roller pins. There are only a few cannelure tools on the market at this time, and the two most popular makes both use this sort of arrangement. One, the C-H tool, operates with the bullet placed vertically, and the other, made by our plant, uses the horizontal rollers. Either is perfectly adequate, as the job is a simple one.

A cannelure has at least one other function of value to the bullet swager. We normally make jacketed bullets in swage dies, since that is a more versatile bullet than the all-lead bullet. But when jackets are

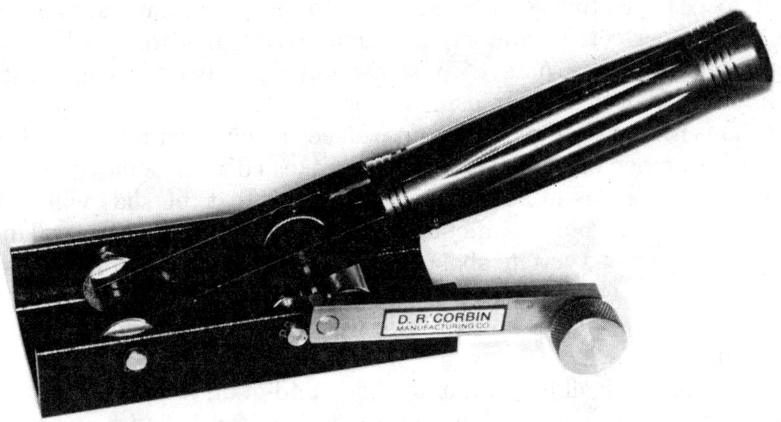

A cannelure tool.

not available, such as is the case at this time for calibers like the .25 ACP using a .250-inch bullet of about 50 grains, an all-lead bullet with a cannelure groove can be made. The swaged pure lead bullet will of course have to be driven at modest speeds. Fortunately, for most pocket pistol calibers and black powder guns, those are the very kind of speeds used.

A pure lead bullet has many advantages. It has better sectional density than those made with lighter materials such as hard cast lead or jacketed bullets, given the same weight and shape. It expands on impact better, at lower speeds, than almost any other suitable bullet material. It also holds together quite well after impact and delivers the most energy to the target. The disadvantages are the leading of the bore, limitations on velocity (basically for the same reason), and deformation in feeding in some guns. When talking about general characteristics, most of us refer to the "average, modern firearm" even if we don't say so. Thus, the problems of velocity limitations and bore leading aren't too important for making swaged bullets out of unjacketed lead, if we mean to use them in low-velocity handguns or black powder rifles.

Feeding problems can be licked, in a number of cases at least, by changing the bullet-seating depth and the conformity of the loaded round's outline with the outline of the cartridge which was designed for that gun. We have tinkered with the use of all sorts of nose shapes in autoloaders, and been able to find a combination of seating depth and nose shape other than the standard round-nose factory autoloader bullet for perfect feeding in just about every kind of self-loading handgun. In fact, some of our test guns fed better with higher reliability factor than when using factory ammo!

This isn't to say that every gun will feed semiwadcutter ammo, but a good number of the ones we actually tested will. Sometimes the powder charge has to be modified slightly so that the slide will kick back harder or softer; sometimes the delay in the cycling will be enough to permit a certain shape of bullet to feed or to jam. But the use of lead bullets of effective shape for good performance is not prohibited entirely just because it takes a little work to get some guns to digest them properly.

The .25 ACP all-lead bullet in about a 48-grain weight, with a 19° Keith-style nose and a rather sharp step between cylinder and nose portion, makes a far more effective bullet than a full-jacketed round nose. These are easy to swage in one simple die and can be fired without lubrication, since the .25 ACP barrel is normally about two inches long, and easy to clean. If a lot of shooting is to be done, then a cannelure groove can be rolled into the bullet, and lubricant put into the groove.

Lubricant can be melted and poured around a lot of such bullets, standing on their bases in a jar lid. The hard lube is then wiped away from the bases, or a metal tube made up of a discarded cartridge case can be used to cut each bullet, cookie-style, out of the mass of lube. Alternately, some lube can be pressed with the fingertips into the groove. It depends on how many bullets are going to be fired. Of course, they can also be lubed in a regular lubricator-sizer such as that used for cast bullets.

Cannelures also serve to identify bullets. For example, a cannelure might identify a group of bullets that used a slightly lighter core from another group of the same jacket length with a heavier core. Match bullets are seldom cannelured; so this is also a good way to keep from mixing up "seconds," useful for hunting or plinking, with the closely matched target bullets. Those which don't quite come up to specs can be set out, cannelured, and thus positively identified for future use.

An irrepressible story comes bubbling up at the thought of the foregoing. It seems there was a match shooter who had tried just about everything to pin down the secret of one-hole groups, and after taking

A power ejector unit, top view.

care of just about all the other variables finally got down to the bullet. He swaged all night long and carefully sorted out each bullet by weight, so that those that matched perfectly went into one pile, and those that were heavier or lighter went into another. He loaded up his cartridges, one by one, using his selected cases and refined techniques, rushed off to the weekend match, and he won!

Naturally, he was only too glad to share the secret with his fellow shooters, and was instrumental in starting most of them home with the concept of perfect bullet weights firmly in mind. But when he got home himself, the first thing he saw on his loading bench was that pile of perfect bullets. He'd loaded the wrong pile. It might point out the advantage of a cannelure, but it points out better that there are many more things under the heavens that affect accuracy than we sometimes imagine. Or, there is less in the bullet after a certain degree of perfection is passed, than in the shooter, and what he thinks he can do.

So much for the cannelure tool. If we have a core mold, or a core cutter, the other tool becomes the accessory. It's nice to have both around, since lead wire turns up even for the man who likes to cast cores, and cutting is faster. The fellow who likes the convenience of cutting wire may find, from time to time, that he is out and can't get more lead wire in the right size for a little while. Then, the core mold saves his day, maybe for quite a while. The prices being reasonably low for these tools, and the times being somewhat hazardous in regard to shortages from time to time in lead wire or bulk lead, having the means to use either kind of material for cores is a good idea.

A very useful accessory is the automatic or power ejector unit. When swaging cores, the amount of time the lead is left in the core swage will determine the refinement of its weight. For very precise work, a little weight change continues to occur in the core as long as it is left in the core swage, from continued flowing of lead out the bleed holes. It may do this for several seconds before it becomes unmeasurable. A good, positive time interval is essential for precise cores. The use of a power ejector (meaning, a device that goes on top of the die, replacing the knock-out rod, or being used with it in some designs, to force out the core using powerful leverage rather than a hammer blow) gives far greater control over the time the core remains in the core swage die.

The use of a leverage-operated ejector unit also cuts down on the inertia resistance of the moving parts, which means less stress on the die parts during ejection. Because it is a smooth, powerful action, instead of a sharp rap, the power ejector's stroke eliminates battering of the top punch head and also makes seated cores and formed bullets

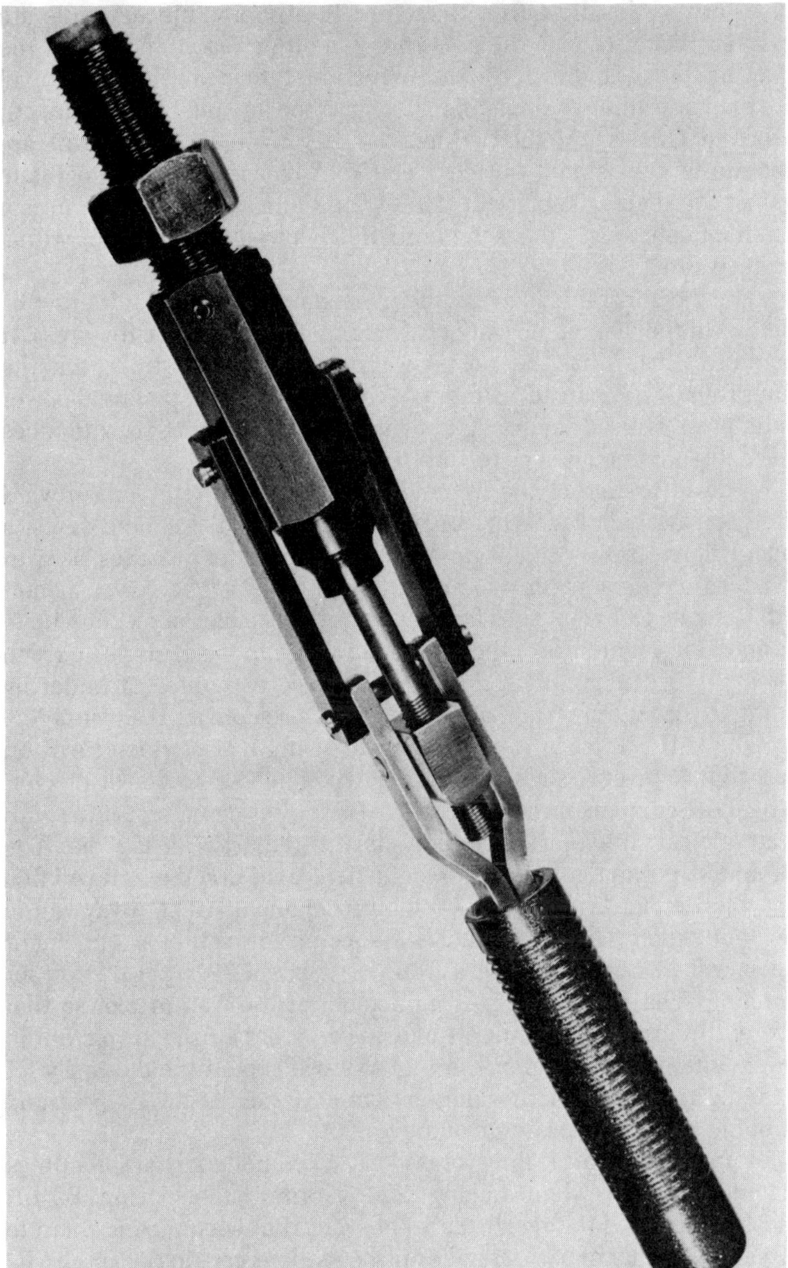

Power ejector unit installed on a die.

"ease" out of the die rather than "jerk" out. Some ejector units are devices which attach to the press ram, and then reach up, around the top of the press, and over the die. When the ram is lowered, the top of the "ejector frame" contacts the knock-out rod and pushes the core out of the die. Others, like the C-H tool and the one made at our plant, are separate lever-operated devices. The C-H tool uses a cam to rotate against the regular knock-out ram top and push it down. Ours uses a system of compound levers very much like a loading press, but with a reversed stroke.

The ejector frame is by far the more simple device, but it must be made to fit the diameter of ram and height of the die above the press. It operates so that it has the most power at the top of the stroke, and the better frames have an adjustment so that nearly all the potential power of the press is used, by fixing the movement so that a screw touches the knock-out knob just as the press starts down.

A disadvantage of the frame type is that each stroke downward will eject the bullet whether or not this is desired. In some cases, a bullet will be left in a die for further forming or other punches, as with advanced techniques for making handgun bullets. The power ejector with separate lever is just as fast, but leaves the choice of ejection up to the operator, regardless of the position of the ram. It can also be used to "force" a special punch shape into the base or nose of a bullet by adding to the ram pressure, although this is a seldom used advantage.

The C-H ejector is very economical, using a disc which turns on an eccentric pivot to press the ram down. The Corbin power ejector, being more complex, is more expensive, and uses a ram of its own with a rather heavy hex-bar body that slips over the top of the die. The Corbin tool doesn't go over the top of a new-style C-H die, but will fit a die which is not larger than 7/8 inch in diameter all the way to the knock-out knob. The C-H tool uses an open frame sort of arrangement which will fit nearly any die as long as it has a 7/8 inch section for clamping. The ram position is adjustable in the Corbin tool so that different internal arrangements of dies can still be used at maximum power. Either the Corbin or the C-H tool will slip off the die easily for changing to use on another die; yet both are held securely by strong clamping action at the bottom of the units.

A method of annealing brass or jacket material is sometimes desired, especially when making .224 or 6mm bullets using .22 LR cases for jackets. Melting the cores of swaged bullets to bond them to the jacket, so that expansion takes place without core loss, can also be a desirable thing. To do this, and to accomplish a number of other tasks such as forced hot air drying of washed cores, bullets, or jackets, and nonswaging gunsmithing tasks like annealing metals and making springs, we have a flameless heat gun. This unit can be used to build a very low-cost heat treatment oven using hot air. The gun runs on 110-volt 60 Hz. household current and produces a hot air stream of 1000°F maximum temperature, at the nozzle.

The room air is mixed with the superheated air and the resultant stream can be any temperature between ambient room temperature and 1000°F, making it possible to do all the annealing and core-bonding without using open flames. A dozen firebricks stacked to form a box will make a nice heat-treatment oven for annealing, and the nozzle can simply be aimed through an opening to play on a metal pan supported within the oven. This is a more even and controllable heat than that of a propane torch, and is a good way to handle 500° to 800°F anneal work. Of course, this is a rather advanced form of accessory and certainly isn't a swaging necessity for most work.

One of the more important accessory items is a good micrometer. We find, however, that a micrometer can also get a person into problems since it isn't the best instrument for measuring bullet dimensions. It is only a rough guide to out-of-round condition and taper. Not only can it barely tell anything about these conditions in a repeatable way, but it does not say if a bullet is deformed in such a manner that it has a warp or triangular shape, provided that the distance from any two opposite sides is the same.

An interesting experiment is to make up a triangular-shaped block of wood with slightly bowed connecting sides and rounded corners, so that this block is the same distance across at any two points on a straight line passing through the center. This block, which looks for all the world as if it would be a bumpy ride, will roll between two boards as perfectly as a round wheel. At least one engine has been marketed using such a rotary piston.

This points out that a micrometer, measuring as it does only the distance between two points opposite each other on a shape between the jaws, can hardly tell us if the bullet is round or not. It can tell us if it isn't, of course, since if it reads differently as we turn it, then it can't be round. But it does not tell, for a certainty, that the bullet is in fact round and not some other equidistant shape. In fact, if the jaws are wider than the deformity, the bullet does not even have to be equal distance from any two points through the center.

A slight burr on the base of the bullet, almost too small to see, will appear as an out-of-round condition. This is very common when the bullet swager first starts to mike his results. After accidentally banging the point-forming bottom punch into the die face a few times, he is making bullets which have a raised portion matching the dent in the base punch. When he mikes his bullets, they appear larger across the portion with the raised part. This, he decides, is an out-of-round bullet and the dies are bad. Yet those dies may be the finest ever built.

Since most bullets will have a slightly larger area just at the base, which is called the "pressure ring" and is a result of the cross-section of the bullet showing a solid copper disc at the base, and a thin ring of copper backed by lead all the rest of the way, plus the fact that pressure is applied at the base and has to overcome friction all the way to the top of the bullet against the die walls, the base will read just slightly larger than the body. In a tubular action rifle, or for most hunting purposes, the pressure ring is not really a desirable thing. It causes the bullet to be a little looser than otherwise in the cartridge neck, if the ring passes through the neck and opens it up a little. In benchrest shooting, it seems to be desired, and of course the bullet is not seated hard in the neck for this application. A drawing die can take all the pressure ring off the base, if desired.

Still, when measuring a bullet, it is best to remember that the base might read a few ten thousandths of an inch larger than the area immediately ahead of the base. A micrometer is a secondary sort of instrument for measurement. It is certainly no standard, since it can easily get out of adjustment. Most shop mikes cannot be relied upon to tell the difference in a half-thousandth of an inch, simply because there is that much or greater wear, dirt, bending, or other faults in the mike that come with use. A bench mike can be kept clean and those with good, larger-size thimbles can read accurately to one ten thousandth for some time.

Optical comparators, electronic comparators, and precision standards are required for truly reading differences as small as one ten thousandth of an inch consistently. Even though micrometers for less than twenty dollars are available that are reported to read 0.001 inch or even 0.0001 inch graduations, there is a lot of difference between having that kind of graduation, and being guaranteed to read linearly across the range to that accuracy. A person can put one-millionth of an inch graduations on his car's speedometer if he wants, and get about the same sort of usefulness. If accuracy of a ten thousandth of an inch is available in the major brands of fairly simple instruments at prices ten times greater than the others, it should tell us something about what kind of accuracy we can expect.

But still, a micrometer is a good thing to have. It gives us a comparison between our bullet and others we make, and those made by other people. It is not any kind of primary standard; if we think it says ".2248 inches," that is no sign at all that the bullet is precisely that size. It might be .2245 inch or .2251 inch, all within a tolerance of less than most shop mikes will handle after any use at all. But since the normal fellow can at least agree that a mike reads to a thousandth of what the other people in the room, using the same mike, think it reads, we do have a good comparison for the nearest thousandth or somewhat less, between any two bullets we make.

A V-way mike, used for measuring three-fluted reamers, is a better kind of instrument although it costs around one hundred dollars or more in the desired precision. It will read more kinds of out-of-roundness and a more precise diameter picture can be achieved. Only a rotating or scanning type of measurement can really tell the whole story, of course, using data from many tiny pin-point measurements from side to side, or in comparison to an accurate center. The bullet spinner, which uses a dial indicator and a pair of cup centers, is a fairly good approach to this.

Dial indicators can read more precisely than a micrometer in terms of a relative size, compared to standards. If a dial indicator is calibrated with a good shop standard, or working standard, and then used to read a similar-sized bullet as it is turned between centers, a reasonably accurate determination can be made of the bullet size, roundness, and taper. One of the problems is that even the medium-priced spinners have enough built-in error and movement in the turning of the bullet that a true picture of all three characteristics cannot be achieved. A good set of bench centers for precision work in the toolroom might well cost six hundred dollars or more. The indicators used with them probably would be electronic, and cost at the very least three hundred dollars more. What chance has a fifty-dollar bullet spinner got against this kind of investment to get the desired results? Still, better to have the fifty-dollar tool than a guess, because the necessary precision is determined only by the desired use of the bullet.

Our friend Ted Smith of the old S.A.S. Dies company once told a story about a fellow who wanted him to make a set of perfect dies. The dies were built, in due time, and shipped, but the man sent one of them back for additional work because it appeared the bullets were very slightly out of round. Ted looked the dies over, couldn't really see any problem, but polished them a little more anyway and sent them back. Then the man was happy. He was so pleased he came by and invited Ted to go shooting with him.

Naturally, Ted went, being a Lifetime Master himself and a person disposed to enjoy both shooting and an opportunity to be elsewhere during working hours if a reasonable excuse presented itself. The fellow drove along to an isolated spot, stopped the car, and got out. He began to gather up some small sticks, and asked Ted if he wouldn't like to gather some, too. Curious, Ted followed his lead. When they both had plenty of sticks, the man got out his rifle, loaded up some of his perfect bullets, and tossed sticks into the air for Ted to shoot! Ted said later that neither one of them would have missed any more sticks if they'd used *square* bullets—or had more fun!

In the final analysis, the most important test of the bullet is what it does at the target. If it shoots into a good tight group, who cares if it is oversize, undersize, lopsided and fore-end-to? What is "size" anyway? If a person thinks that all .224 rifles have a .22400-inch barrels, he is at best a supreme optimist—some might say an idiot. We would like to think that barrels are all perfect, since they aren't too easy to measure. After all, what is more convenient than a standard that no one can dispute because no one can check it?

If a barrel is slugged, which is what we call shoving a lead plug down the bore and then measuring the resulting size of the plug, it will read somewhere between the tightest spot in the barrel and the amount of force that was necessary to shove the slug on out past that tight spot. That's all. It's better than nothing. It's no standard. There is a better way, still not really a standard. The use of an air gauge will permit continual plotting of the diameter of the bore, within reasonable limits of roundness, as the gauge is pulled through.

We have about fifteen air gauge columns at the plant, all very nice commercial units, that seldom get a lot of use except in comparative work. The bullet that comes from a die is never the same size exactly as the hole in the die. And the best bullet for a given barrel probably can't be determined precisely, to the degree some would have us believe, simply by getting an idea of the size of the bore. The bullet swager can determine better than anyone which diameter fits and shoots best in his rifle, simply by using more or less lube, more or less pressure, and making slightly larger or smaller bullets all of the same shape and weight to test. What he gets for best results probably won't do in any other rifle, but of course that is exactly the point.

To specify a bullet of exactly 0.3085-inch diameter for a certain rifle only makes sense if a person has found that a bullet 0.0005 inch smaller or larger will not group well in his gun. Unless we have tried it, how can we possibly say what the best diameter will be for any given barrel? The degree of accuracy specified is completely out of reason with the certainty of our standard, in some instances. Many times a set of dies will be returned to be enlarged a few ten thousandths of an inch, and in most cases it is because a micrometer said that the bullets needed to be bigger, not because the bullets didn't perform. Sometimes a comparison using a standard shipped back and forth between plant and customer will show that what one micrometer says is almost a thousandth of an inch different from what another one might.

It is almost a necessity, to really do what is expected in these cases, that the customer's micrometer be sent along with the dies so that the bullets will read what his micrometer would like them to be, after all is finished. On the other hand, if a bullet shows bad out-of-round condition even on a micrometer, and it isn't just on the base but all the way up the sides, then in all likelihood the bullets won't shoot well in any gun. We have to rely upon good judgment of sizes when we get into these very small tolerances. Usually, a bullet which is a thousandth or more one way or the other of the accepted standard will not be as good a performer as it should be, but not always. On the other hand, a few ten thousandths of an inch larger or smaller, as long as it is consistent, will usually not make any difference in the accuracy.

A good scale is a nice accessory item for bullet swaging. Actually, the weight of the bullet has less to do with its accuracy than the proper seating of the core, but still the weights ought to be reasonably close. If a person can get no more than half-grain variations on his bullets in a .224 caliber, he is getting very good bullets as far as weight consistency. Actually, the weight variation can be brought to half that with care. But the accuracy isn't as affected in most cases by weight as by concentricity and diameter consistency. A technique to ensure better concentricity of the base is to turn the bottom punch in the point-forming die set around and around as the bullet is swaged several times, which tends to even out any slight deformation brought about by the base punch not being perfectly flat. Sometimes a slightly cupped base is applied, as it is easier to make a good punch shape in the lathe and make it concentric with rounded corners than by surface-grinding the end; although there is slightly more cost in the round type, as it requires a little more handling, and setup.

Other than these, there are rather few items which are exclusively the province of bullet swaging accessory use. In the field of redrawing jackets, there are of course the drawing dies and punches, and the pinch-trim dies. But these are rather simple devices, and are somewhat advanced-use items anyway. Most bullet swagers can obtain jackets in the caliber they want, though not always in the special lengths, and the redrawing dies (which Corbin calls the "Jacket-

Maker" sets) and trim dies permit the use of larger-caliber jackets of standard length for smaller calibers of longer than normal size. A .750-inch .243 caliber jacket, for example, can draw to as much as a 1-inch long .224 caliber jacket. These operate in the home reloading press and require relatively little force, although they are limited to about one caliber reduction in size per die and punch set. The scales, reloading press, even some of the casting equipment, can all be used for primary purposes other than swaging. And this is one of the reasons it is easy to get started in bullet swaging without a really large investment; swaging is for handloaders, and handloaders have most of the things they need to start.

Chapter 25

Avoiding Damage To Swage Dies

That's a mighty negative title! But one of the most important things to remember about swaging is that the pressures are high, the forces great, and the tolerances extremely small—all of which boils down to great potential for hurting something unless a good understanding of how to avoid damage is developed.

Actually, there isn't any reason why a person who can drive a car without running over too many pedestrians can't run a swaging press without any problems. But all too often, in a fit of frustration, we might unthinkingly yank on the press handle when we ought not. A vast flow of damaged dies and punches streams into the plant with some very standard and predictable (avoidable) problems every year. This chapter will cover them in the hopes of slowing down that stream to a trickle.

Most of the problems result from one of two operator errors: too much force, or misalignment of the die and punch. It has been pointed out in a former chapter how easily tons of pressure can be developed inside the small bore of a swage die by overzealous application of weight to the press handle. Misalignment of the punch and die usually results in a damaged punch, which is not a really bad problem at all, except that until the punch is repaired or replaced the bullets probably won't be worth shooting.

In core swaging, there can be a particularly pesky problem with punch alignment because the fit of the punch to the core swage die has got to be so very fine. Using a regular slotted-ram loading press means that there is only a certain small degree of "float" to the punch. It has a little ability to scoot around inside that slot with the tolerance of the matching "button" being the source of this ability. (In machine work, we call this "working tolerance" while our critics might call it "slop." But here it helps, especially since not every press has exactly the same size of slot in the ram.)

Now, suppose that the bullet swager is just finishing up a nice big run of regular reloadings with his press, and has got primer residue, powder flakes, shavings off the rims of the brass cases, and other assorted crud mixed in with lots of lubricant and worked into every crevice and slot it can find in the press. He goes to swage some cores. The die is put into the press, and the little button-fitted punch slipped into the ram slot. But instead of slipping around so it can align with the hole in the die, the very back of the ram slot holds this punch button a little to one side. Why? Because of the crud still in the slot.

Now, the core swage die uses the smallest punches and die holes in the set for that particular caliber. If any punch is going to get bent, it probably will be the core swage punch. Not only that, but the core swage operation is the first one in swaging bullets. Usually, the process is as follows: The bottom edge of the punch button keeps the punch just a little out of line with the die hole, due to a powder flake or primer residue jammed between the ram slot and the punch button. But it is not enough to keep the die from lining up on the end of the punch. After all, the punch can tip just a little.

As pressure is applied, the punch presses farther into the die hole and the mouth of the tool steel die gets closer to the rigidly held button. The button would like to slip over so that the punch is straight in the die hole, but it can't. Continued force runs the die mouth against the punch's side, tearing a deep gash into the side of the punch. How in the world can that kind of a tear happen, asks the swager, but we can see it isn't hard at all if the press is not kept clean.

Swaging presses which are not used for reloading don't usually have this trouble. In fact, with floating punch alignment instead of the slotted-ram system, there is seldom any evidence of this kind of damage. But even a hard-used reloading press can be cleaned with a swab stick in the ram slot before using swage dies and punches. There isn't any problem if the press slot is clean.

In core seating, the next step of swaging a bullet, there are a couple of common problems with alignment. First, the jacket may not be put on the punch correctly. It might seem strange that this simple step can be done wrong. But when a person is making a long run of bullets and gets in too much of a hurry, anything is possible. The most common problem is failure to get the jacket over the punch. The jacket and core are rammed into the die mouth, not usually hurting that tough die any, but throwing a strain on the punch and bending it. The second alignment problem is simply not getting the punch under the hole in the die.

With the jacket and core on the end of the punch, there isn't much trouble with damage to the end of the core-seating punch. But when setting up the die, it is common to pull down the punch several times and eject the core and jacket, or simply pull down the punch and then push it back in again to see if a tiny bit more adjustment might feel better. That's when the punch gets bumped over to one side and then whacked into the face of the die instead of going in the hole. This spoils the nice sharp edge of the punch so it no longer fits perfectly at the junction of the lead and the jacket wall. Sometimes it raises a burr so that the punch just won't go back into the jacket properly. The repair is easy; a jig made to hold the punch straight over a grinder and rotate it is needed to make the punch tip flat and even again. It is standard practice to regrind the faces of punches any time a die is sent in for inspection or repair; usually, they need it.

A burr on the end of the core-seating punch is only serious if it cuts the jacket or keeps the punch from fitting properly. But a small dent of any kind in the edge of the point-forming bottom punch will transfer a similar reversed deformity to the base of the bullet. That's bad. The point-forming punch (the one that pushes the seated core and jacket into the point-forming die) needs to be perfectly flat and smooth. It can just as easily be bumped into the edge of the die as any other punch. That is, if we get in too much of a hurry.

So many times a die set will be returned for "adjustment" with a complaint of poor bullet grouping, only to find that a five-second pass over the grinder will take off a bad dent, square up the bullet base, and produce match-quality results. All too often the beginning swager damaged the punch before he even made the first bullet, running the punch up and down into the die to check it out. And as a result, he doesn't believe that the dies are capable of making good bullets!

The base of a bullet for either .224 or 6mm rifles (which are the calibers made for swaging in a reloading press, larger ones being suited only for positive alignment, higher-pressure presses using solid rams) can be flat, or it can be dished inward, and excellent groups made. The dished base is a result of using a slightly domed punch in both the top of the core-seating die and the bottom for the point-forming die. The dome must be very even and concentric, but it does tend to put pressure first in the very center of the jacket, and then toward the edge. This will start the lead swaging from the center toward the walls, forcing any air out of the jacket base. The technique also works better for using 6mm bullets made with .22 LR cases, as these jackets are far undersized and need the center-to-edge application of pressure to form an even base.

While a domed punch is a little harder to make, it is worth the difference with slotted-ram presses. Some slight guidance of the punch into the die mouth results, and damage is reduced from accidental misalignments. However, it isn't too hard to develop a safe, slow cadence to swaging bullets, even in a long run. And the technique of holding the bullet gently between thumb and forefinger while the punch edge is felt easily, slowly passing the mouth of the die (making sure no finger gets into the works) will eliminate any punch damage.

In swaging all-lead, half-jacket, or three-quarter-jacket handgun bullets, using the straight-wall die and a nose-forming punch, it is all too easy to let the nose-forming punch, with its thin-walled edges, slip over and ram into the die face. It's avoidable, but it happens. The punch is usually destroyed by having its edge folded inward. If this should happen, it is best to forget about repairs and just replace the punch. In fact, it ought to be sent back right then, not used for one more bullet. The possibility of sticking it in the die is great if the end is warped and bent.

The cavity in the end of a handgun nose-forming punch is deep and the walls of the punch are thin. Usually, they are between 0.015 and 0.025 inch thick. That leaves a nice little semiwadcutter effect on the bullet nose, just above the end of the jacket. But the thin walls and deep cavity mean that the punch can expand in diameter. In a few dozen bullets, the punch may be a tight fit for the die. This is a little like using a break-in period to fit the piston rings to the cylinder of a car. For the first dozen bullets or so, depending on the thickness and depth of the cavity, there may be some lead extruded past the punch edge. This ought to be scraped off with the fingernail or a wood stick (not metal) to allow the punch to expand until it is a perfect, although tight, fit in the die.

A little fin of lead on the edge of the bullet is the result of this break-in time, but it is easily folded over or scraped off with the fingernail. It will go away after the punch fits itself to the die. To avoid scoring and sticking, the punch should be wiped with a film of lubricant along with the bullet nose, every time a bullet is swaged. This is very easy to do, since lube is normally applied with the fingertips and those same digits are going to be hovering around the punch tip anyway, ready to put another core and jacket on it.

The punch can be covered with burnish marks and not hurt a thing; these are very small hills and valleys in the noncritical lower part of the punch, that appear only when very precise fits are being approached. They spoil a mirrorlike shine, of course, but only the very tip of the bullet-forming punch really needs to fit snugly to seal off the pressure. A dull, burnished area, especially if repeated around the punch diameter, is not damage. But a deep, burr-raising score mark is.

Quite often, the lower part of the punch will be turned slightly smaller than the last quarter inch or so near the tip, just to ensure a precise fit won't be spoiled by expansion in the lower areas with continued use. Too much area turned exactly to fit the die bore just increases the friction without contributing to a better bullet. The same effect of dirt in the ram slot as affects a core swage punch can force a handgun nose-forming punch into contact with the die face at midway down its length, and a little relief on these larger punches helps eliminate this effect.

Using an oversized jacket, or a dirty one, or using a core and jacket swaged in a similar size or larger core-seating die in a point-forming die, can all cause damage to punches by requiring far too much ejection pressure. The components need to be clean, and they need to be smaller than any die into which they are about to be pushed. A jacket that is already .2235 inch in diameter just cannot be expected to fit into a core-seating die that is .2230 inch in bore. It will go in, of course, since a mere 0.0005 inch is a fairly easy force fit even in steel parts. The punches and buttons for smaller calibers are often a 0.001 inch or greater force fit even with tool steels. But the force to get it out again can be damaging to the long, thin top punch, which we know has to be as long as the die at least.

In point forming, if the bullet or seated core and jacket is so large that it takes considerable pressure to fit it into the die, then there is almost no hope that the tiny ejection pin can push it out. Instead, the pin will penetrate the bullet, meaning a return to the plant for service. Actually, it usually isn't too hard to get a stuck bullet out of a point-forming die, and it's a free service to owners of the dies. Sometimes it gets to be a case of being cheaper to provide a new die than to get a badly jammed bullet out, though.

The process for getting a stuck bullet out of a point-forming die, especially if the bullet is stuck only due to lack of lube or slightly

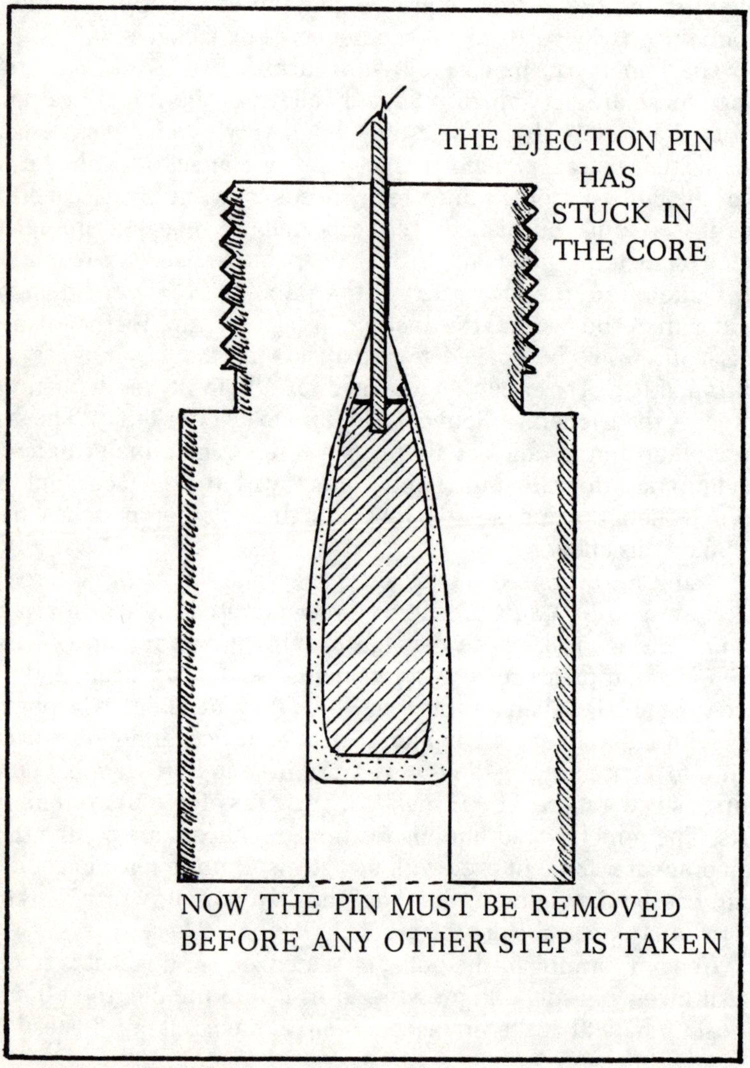

A bullet stuck on the ejection pin.

oversized jacket, is as follows: First, remove the die from any adapter it may be fitted with, and pull out the ejection pin. Drop a bit of lub into the ejection pin hole, and put the whole thing back together again. Push the bullet just a touch farther. This can easily be overdone, but the purpose is to close the hole that the ejection pin made in the bullet and hopefully bring the jacket around so that the pin bears on it. This is for rifle bullets only, not lead-pointed handgun bullets.

The bullet may eject on the next try. If it doesn't, repeat the above but do not put the die back together again. Instead, cut a long piece of spring wire the same size as the ejection pin (not an Allen wrench, for heaven's sake!) but about half again as long as the entire die. Put the die between lead jaws of a stout vise, and tap the end of the spring wire with a mallet to drive the bullet out. If it doesn't drive out, the next step is just to let nature take its course and drive the spring wire right through the bullet and out the base.

The wire will have made a hole through bullet and jacket, and let daylight through the entire die, after it is pulled through with stout pliers. An "Easy-Out" extractor or a hard tapered tap small enough so that its teeth will not break through the jacket and touch the die walls can be turned into the hole from the big end of the die. A clamp can be firmly set around this tool, and the bullet pulled out by application of strong leverage between the die face and the clamp. Or so we hope.

If this fails, in all probability a jacket and a tiny bit of lead will be left in the die. Now it is time to let precision methods reign. The die can be chucked up and a precision boring tool or grinder used to make a groove which can be used to pull the jacket from the die, or various chemical means can be used to dissolve the lead and copper out of the die without harming the steel. Probably, unless the die is very costly, the best answer is to chuck it and go for another one. That's what the usual plant procedure is on lower-cost dies. Fine, labor-of-love benchrest dies are worth extra efforts to save.

The bullet which cannot be removed economically, in a lower-cost die, is usually the result of trying to swage down a commercial bullet. Never try to swage down a commercial bullet in one of these dies. The tip shape can be changed on open-tip or soft-point bullets without much problem, but a 6.5mm bullet just won't make a 6mm super heavy experimental bullet in a one-step swager! Likewise, a solid jacketed-point bullet can be made in a simple low-cost die set with the base of the jacket swaged into the point of the bullet; but if it sticks, it's stuck. Normally, the reason it sticks is because enough pressure was applied to form the very thick base portion of a tapered-jacket bullet into a nice sharp spitzer. The trouble is, no one will ever get to see it as it is doomed to stay in the die forever.

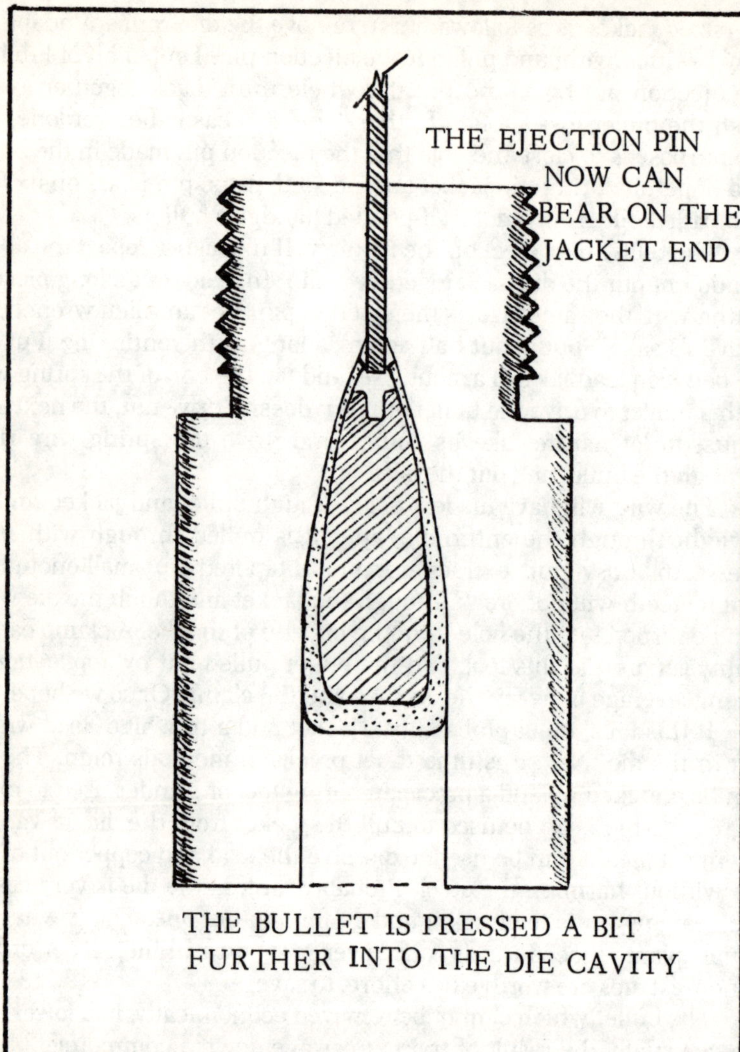

Ejection pin free and bearing on the jacket.

When great amounts of pressure are used, it is just as well that the bullets don't come out again. The die most likely will have acquired a swollen midsection. A pot belly in the center of a precision swage die means that the bullets which might happen to be made in that die have to be resized by the mouth of the die when they are ejected. That is, if they can be ejected at all.

While having nothing to do with swage dies in particular, it is best to touch on one other damage problem. It isn't necessary or desirable to use a pair of "water-pump" pliers to tighten the die while holding the locking nut with a giant crescent wrench. If the die tends to work out of adjustment, place two locking-thickness 7/8 × 14 T nuts on it and run them down against each other while holding the die with the fingers only. If a die turns so hard in the press that pliers are needed, then either the threads of the press have powder grains and grit in them or the die is fitted to a defective or burred holder and should be returned for replacement. Grasping the top of a die adapter with pliers is bad not only for looks but because it can raise burrs that prevent the use of a power ejector unit—a very desirable accessory item that clamps over the top of the die to eliminate the use of a mallet in ejecting bullets.

Undoubtedly, a majority of the users of bullet swage equipment, or those contemplating such activities, would never consider such obvious abuse. But imagine the feelings of the diemaker who worked late night after night on a particularly critical set of dies, lovingly polishing a bit here and a spot there, testing, fitting, tinkering until the set produces bullets like golden jewels to his perhaps cynical, work-hardened sense of beauty. Imagine him wrapping each punch and die in cotton packing, carefully placing each child of his labor in layers of protective wraps and holding the last part a few minutes to relish, in the fading evening light, those rich gleams that play along the perfect surfaces between his fingertips. Imagine, if you will, his lack of diplomatic poise when that same package is returned, bearing the scars of tong and pincers, with the cryptic note that "I think these dies need checking; they don't make good bullets any more. P. S. I filed off the punches but I think there's still a burr in the die."

Chapter 26

When The Bullets Won't Shoot

By nature, the person who buys swaging equipment is usually interested in better performance than he can obtain from factory bullets—sometimes at a lower price, or sometimes in sizes and weights that can't be purchased. Sooner or later, the bullet swager wants to find out just how his bullets stack up against precision factory bullets. Here a wide range of performance is acceptable, depending on the person, the gun, and the purpose. Before judging either the technique or the equipment, one must have goals that are practical and consistent with the firearm, purpose, and shooter's skill firmly in mind.

A factory bullet and a home-swaged bullet start out with equal chances of being knocked off course by factors other than those of the bullet's quality. Even if the dies are perfect, the technique the most consistent possible, and the bullet appears to be made to perfect standards, there is no guarantee that a box of such bullets will group well in any rifle, under any circumstances. It is always possible that the load of powder behind a given weight, shape, jacket hardness, or bearing surface of bullet will fail to extract the accuracy possible in a given barrel. One of the virtues of swaging is that it gives the handloader complete control over weight and a wide control over other factors of design. If the first bullet design doesn't shoot, change one factor at a time until the results improve.

Unfortunately, many shooters have picked up the idea from magazine articles and talk around the ranges by shooters who should know better, that just using benchrest-grade bullets will automatically turn a reasonably good rifle into a superperformer. Not so. There is a great deal more than the quality of the bullet behind the group, although good bullets are a big factor. The gun and the shooter have to be capable of the expected performance before one can start troubleshooting the bullets or blaming them for lack of achievement. Reading a list of what the winners shoot, ordering a set of the most popular dies or bullets, and then expecting to make the same groups is not very realistic. It might be, if the shooter already made those kinds of groups consistently, and his wind doping, loading, and other techniques of shooting are well under control. If he has done it before, he can expect that his hand-swaged bullets will also let him do it, once he gets them tuned to his gun and load.

Again, the shooter may not feel badly at all to get a one-inch group at 100 yards if he is shooting a Mini-14 or other fun gun. For him, a lower-priced set of dies costing less than a day's wages for a skilled machinist (this is easier to keep in relation to the times and the inflation rate than dollar amounts) would be more than adequate. Such dies should make bullets that have roughly a minimum quality as follows: shank taper running to not more than a thousandth on a .224 caliber bullet, roundness not out by more than about a half-thousandth, diameter range of not more than .2245 in the shank and not less than .2235, a pressure ring somewhere around two-ten thousandths larger than the shank, and only slight burnish marks at most around the ogive, no deep rings or circular impressions. None of these things would affect the performance enough for the average casual shooter to notice, and the bullets would in general be cheap enough and good enough at the same time to make the swaging operation well worth the cost.

It is possible to offer dies that cost less than one day's wages for a skilled machinist because most people are perfectly happy with tolerances that fit within the above category. If tighter control were required, then the extra time involved would push the cost of the dies into the benchrest class. In considering the purchase and testing of swaging dies, it is important to recognize just what class of performance is really desired, and what quality and price of die is being purchased. Lower-priced dies very often will make bullets that group nearly as well as the most costly benchrest dies. In fact, it is hard to tell the difference in quality in many cases. But if one really intends to depend on getting this kind of quality, then only by specifically ordering it can it be guaranteed. The price then is more like a week's wages for a skilled machinist.

Even with the most costly dies, there is a strong and unfortunate tendency for some precision shooters (or would-be shooters) to reject a bullet design or a die set based on factors other than the group. The final test—the only meaningful test—of a set of dies ordered for production of benchrest bullets is the groups that they produce. Experience at making dies for benchrest work has proven that most of the rejections or returns come from shooters who have not even tested the bullets in a gun! The most common of the reasons for rejection is some minor cosmetic feature of a die set, having nothing at all to do with the bullet quality. Another common complaint might be that the bullet produced—the only style attempted, quite often—did not look like a certain factory bullet or had a burnish mark or a tip opening different from some standard bullet. A shooter really owes it to himself to fire a few bullets after having gone to all the trouble of making them, and then do his troubleshooting on the basis of the group. The cosmetic factors can be changed as experience is gained in swaging. Many are not die faults, after all.

But should a set of dies produce bullets which do not group well, such as making two-inch groups at 100 yards in a gun which fires factory loads and regularly gets one-inch groups, what can be done? First, the bullets should be checked for uneven bases. This is a very common problem, caused by striking the bottom punch edge against the mouth of the die. The edge of the punch folds over or acquires a dent. This dent or fold allows part of the jacket material at the bullet base to extrude past the baseline. The effect is a bullet which will measure on a micrometer as if it were out of round. The uneven base will allow gas pressure to upset the bullet from its path as the bullet leaves the muzzle. Because beginning swagers often are not careful with the punch and die alignment, this problem shows up initially and is thought to be a die fault. Checking the punch edges for burrs will usually tell the story.

A second problem is the use of rimfire jackets for 6mm bullets, where the bullets are driven to higher speeds than recommended. In spite of warnings that this will mean fliers and inaccurate bullets, it happens frequently. The problem is that a standard .22 long rifle bullet jacket (meaning, a jacket made from a .22 LR case) is only about 0.705 inch long. This is ideal for a 53-grain .224 bullet, but to make a

60–65-grain .243 caliber bullet a large lead tip will have to be exposed. These bullets shoot well enough if the velocity is held down, so that the lead tip does not flow back onto the shank or deform against the air during flight. Some very fine groups have been sent back to the diemakers by people who have no reason to do anything but tell the truth. In each case, the velocity was well below maximum for a 6mm rifle, generally less than 3,000 fps.

When using thin jackets, and experiencing poor accuracy, the first thing one should do is reduce the powder charge and work down to lower speeds. At speeds in the 3,500 fps and higher range, the thin jackets may not be holding together well in flight. There is no hard and fast rule about maximum speed. Some rifles may not cut the jacket enough to start it opening, while others with sharper rifling or a tighter spot in the bore might weaken the thin jacket so it opens sooner, at lower speed. The rate of twist is very important as this is the primary force that throws the jacket apart in flight. If bullets are tumbling, as evidenced by keyhole marks in the target instead of round holes, reducing the velocity is a good step toward working out the cause of the problem.

The jacket should be selected so that the lead core does not come to rest, after seating, much below the two-thirds point on the ogive. That is, a bullet should not be used that has more than about a third of the ogive open, without any lead behind it. This is not a hard and fast rule, but experience shows that such a combination leaves too little support for the jacket tip, and either the bullet will not form well in the point forming die, or it will not shoot well. Ejection problems can result from leaving too little lead behind the ogive. If the ejection pin is smaller than the desired tip opening at the jacket end, and there is very little core weight for the jacket length, this means that there is nothing for the ejection pin to press against until nearly all of its stroke is used up. When lighter bullets are needed, a shorter core should be used with a jacket that matches; not over a third of empty ogive should be permitted.

Very rarely will a bullet not shoot well because it is over or under a certain diameter by a few ten thousandths of an inch. Usually the problem lies elsewhere. If a bullet is very much undersized, it probably will not be very accurate. The amount at which trouble starts to show up depends on the gun and the precision expected, but generally speaking, a half-thousandth of an inch is getting to be the limit. When a bullet is a full thousandth undersize, it can probably be expected to have problems, and even the casual shooter may notice. A bullet that is oversize will raise pressures on the hot loads, but normal loads generally have enough tolerance to mask the difference in a bullet a thousandth oversize. Groups may not even suffer for the casual shooter; a benchrest shooter would notice. The same holds true of slight burnish marks on a bullet ogive or shank; these have no effect on accuracy, but are a result of diamond still in the dies from lapping (it comes out as the dies are used) or sometimes a lack of sufficient lubrication on the bullet.

A very common problem with undersized bullets is the use of the improper size of core-seating punch with the jacket selected. The core-seating punch must fit properly. It shouldn't allow a great deal of lead bleedby, and it shouldn't scrape the jacket walls. Since jackets are made with different wall thicknesses, the core-seating punch is the one part of the swaging set which depends entirely on the jacket and not on the bullet diameter. Jackets for rifle bullets taper inside; so a narrow range of weights is all that a given punch can handle before a different diameter is needed. And a problem with fliers and inaccurate bullets also results when the shooter fails to change the core-seating punch, and makes the ogives using the same punch as he used to seat the cores. The core-seating punch is smaller than the bore of the core-seating die (except when making lead-tip bullets, of course). Naturally, it is even smaller than the bore of the point-forming die. The result is a depression in the base of the bullet, extruding jacket material around the punch. It may not be centered in the bullet base, unlike a proper cup base. Also, the sharp edges of the punch cut and weaken the jacket. An examination of the bullet base will point out this error.

Generally speaking, bullets which do not shoot have obvious physical problems that can be cured by the handloader or else they are not well matched to the load and gun. A close examination of the bullets, especially of the base, will show up most of the common physical errors. Experimenting with the velocity will usually take care of the other problems. When a bullet still does not group as well as other bullets of a similar kind, fired alternately—that is, not compared to what the gun did five years ago with a similar bullet, but fired against the bullet used for a standard on the same day—then by all means the shooter should send samples of cores, seated cores, empty jackets, and finished bullets to the diemaker. By examining all the components and each stage of the operations on them, the diemaker may be able to determine a problem either with the components or the dies, such as a mismatch between the core-seat and point-form dies or excessive taper in the point-form die.

The important thing to keep in mind is the level of performance possible with the gun and load and shooter, and the level of performance which can be expected from the price range of the equipment. If it is recommended for benchrest work, it should be capable of benchrest grouping bullets. If it is not, then at least it should be capable of making bullets as accurate as the standard factory projectiles. But in no case is it assured that the first bullet out of the dies will perform accurately; not until the swager has learned all the basics and gotten the bullet to match the load and rifle should this be expected. The shooter who makes good groups the first time with his hand-swaged bullets is well on his way to understanding the process, or else, he's just lucky!

Chapter 27

Bullet Swage Troubleshooting

Problems with swaging can be either the fault of the operator, or of the equipment and components. By far the most common problems are caused by operator error; the next most common are caused by faults in component selection. Actual die and punch problems are fairly rare unless caused by mismatching or misuse. It is difficult to produce a die set without producing a number of bullets for testing, which means that usually a die set will have produced good bullets before it was shipped. If it does not do so upon arrival, chances are great that either component selection or operation is at fault.

One of the more common problems is hard ejection of either the seated core and jacket, or of the finished bullet. Hard ejection can come from oversized components—especially oversized, out-of-round, or dirty jackets. In any swage die, the component that is put into the die must be an easy slip fit before swaging. It must be smaller than the component diameter after swaging. In no case should a jacket or a bullet be put into a die with a cavity smaller than the jacket or bullet. Swaging is a process of applying pressure to increase the diameter of a component, as used in making bullets. When a component needs reducing in diameter, then an open-ended drawing die is used.

A second cause of hard ejection is improper lubrication. Each step in swaging which involves the use of a die should be preceded by application of a thin film of lubricant on the outside of the component being swaged. Usually, just rolling a drop of the proper lubricant between the thumb and forefinger and then rolling the component between these digits will do the job. Ordinary case sizing lubricant should never be used; only the recommended high-pressure swaging lubricant is adequate. A mixture of half anhydrous lanolin and half castor oil, or the synthetic Draw Die Lube marketed for swaging and drawing applications is recommended. Lack of lubrication will normally cause hard ejection and stuck bullets in the point-forming die. It contributes to wear in all dies, and should be avoided.

If a seated core has the lead far below the end of the jacket, it is possible that when this bullet is formed the ejection pin in the point-forming die will enter the open end of the jacket and have no lead to bear against. Rather than hard ejection, the bullet will not be moved far enough to allow any ejection. If this should occur, the bullet can be pressed a bit farther into the die to bring the jacket mouth more nearly closed. Then the ejection pin can bear against the edge of the jacket. This problem also occurs with hollow-point, lead-tip bullets. The large cavity means that the ejection pin has nothing against which to press. Most of the ejection stroke will be used up just moving the pin through the open cavity in the bullet,so that when finally the pin touches the bottom of the cavity, there is not enough stroke left to press the bullet from the die. There are several solutions, but generally a shorter cavity, or a different die and ejection punch are required. Sometimes, the pressure required to eject a long-ogive bullet, such as a 6 or 7 caliber spitzer, is suitable for a small ejection pressing on the jacket ends, as with a normal open-tip spitzer bullet. But when the same die is used to make a flat-tip or hollow-point, lead-tip bullet (by pressing the seated core only a short way into the die rather than forming up all of the ogive), the pressure to eject may be greater than the pressure that the small ejection pin can put against the soft lead core and not penetrate it.

Stuck bullets usually occur as a result of an ejection pin pressing into the lead core rather than moving the bullet out. In some cases, there is simply no way to make the desired bullet with the particular type of die and ejection pin and jacket length. A change in components, or in the design of the bullet, may be needed. Usually a slight change in the ogive length or the cavity depth will permit normal ejection from most dies. A change in brand of jackets may be required if persistent sticking occurs and no other cause is found. When using rimfire cases drawn to make .224 or 6mm jackets, a stuck bullet can result from the bottom of the case (now drawn into a jacket) popping out. This is simply a fault of the particular jacket, caused by a weak head, and not avoidable. It is necessary to know how to remove the various kinds of stuck bullets without harming the die in the process.

Usually, a stuck bullet simply has more of a grip on the die walls than the pressure required for the ejection pin to penetrate the lead. By first getting the ejection pin unstuck from the lead and then pushing the bullet a little farther into the die cavity, the jacket ends can be brought closer together so that the ejection pin can bear on them upon the next try at ejection. A particularly good way to do this is to cut a very short piece of lead core and dip it thoroughly in lubricant, then push it into the die right behind the stuck bullet. The lead will seal the bore of the die, trap lubricant between it and the base of the stuck bullet, and even force some lubricant alongside the bullet. Usually, the bullet can be ejected on the next try. If this does not work, then it is best to send the die back to the factory for service, and removal of the bullet. Drilling out the bullet is the same thing as throwing the die away in 90 percent of the cases.

In every case, it is necessary to remove the ejection pin from the bullet before any further work can be done. If the bullet is pushed farther into the die with the ejection pin still stuck in the bullet tip, then the jacket will only be forced harder against the pin and removal will be made more difficult. Whenever a bullet sticks, the first thing to do is quit pressing on it, the next thing is to make sure that the ejection pin is free, and then steps can be taken to remove the bullet. The use of a lead slug with plenty of lube on it works in a point-forming die whenever the bullet is still in one piece and the core has not been pushed right through the jacket bottom. Whenever the ejection pin is driven all the way through the bullet, and the bottom of the jacket is broken out, there is a good chance that the jacket will still be left in the die bore even if the core is removed.

Also, it sometimes happens that a handloader will mistakenly put an empty jacket into the point-forming die. If this happens, there is no lead on which to press for ejection. Usually, forcing the empty jacket up into the tip of the die will fold it up enough so that the ejection pin can bear on the crumpled end of the jacket. If this fails, the technique most successful is to use a regular lead core, and plenty of lubricant, and simply force the jacket to fold up into a compact mass of metal at the die tip. The core and jacket can then be ejected as a unit. If the base of the jacket breaks out on the first attempt at ejection, then a different situation exists.

Whenever a jacket with the base broken out is left in the die, pressing a lead core behind it will normally only fill it again with a core, which will again be ejected without pulling the jacket along. The safest way to achieve ejection of the stuck jacket is to use another, smaller-diameter bullet. The smaller-caliber bullet can be pressed right up inside the stuck jacket, gripping the jacket snugly enough to allow both to be ejected at the same time. This is a good technique for the larger calibers, but for .224 and 6mm the correct bullet size is 0.17 and .20 caliber. These may not be available. In this case, a seated core with jacket can be pushed into the die backward, with the jacket base toward the stuck jacket. The solid metal will push the stuck jacket higher into the die, folding up the nose, and permitting the ejection pin to have something solid against which to push.

If these techniques fail, it is best to send the die back to the factory for service. At the factory, the technique is to drill a small hole in the base of the stuck bullet and insert a proper-sized self-tapping screw or a metal tap. With a stuck jacket, the proper-sized tap is screwed into the jacket so that the threads just cut into the jacket walls but not into the die. These seem simple enough, but the highest accuracy in centering is required. The slightest error will put a deep mark into the die cavity and ruin it. If this happens at the factory, then a new die is furnished but if it happens in the handloader's hands, it has to be his responsibility. This is a good enough reason to send problem dies and stuck bullets back for service rather than attempting to drill them out. No reasonable person could expect the factory to replace a die that had drill or tap marks in the bore.

On the subject of marks and defects in die bores, it is usually very easy to determine if a die has had damage from improper bullet removal in the field. All the processes of making the die involve rotary operations, so that any defects in the die would be made at the factory in the form of rings or concentric circular marks in the die bore. Not one of the processes of reaming, honing, lapping, and polishing involves any tooling that could leave vertical marks in the bore, or marks on one side of the die only. The machinery is extremely precise in regard to concentricity, and most marks that are left as factory defects would be very concentric rings from a chip under a reamer edge that would not be completely removed in the final sizing of the die cavity.

If a die is sent to the factory for warranty with a gouge or scrape mark, or a nonconcentric drill or tap mark along one edge of the cavity, it is a sure sign that the die has been tampered with in the field, usually to remove a stuck bullet. Sometimes a machinist or hobbyist will decide to "improve" a die by changing the ogive, enlarging it at the base, or some other operation that might seem simple enough if it were not for the extreme requirements of finish and shape control. These modifications almost always result in a ruined die. The years and years of experience in lapping dies can seldom be duplicated in a first chance attempt. There is a good reason why almost no one is in the swaging die field compared to other handloading products; if it were easy to do, everyone would be tempted to do it. Hobby diemaking can be fun, but seldom is it wise to attempt it on a factory set of dies no matter how much experience one has at machine work. One should start from scratch and make dies from start to finish; it is easier than modifying a finished, hardened, and tempered die.

A final technique for removing stubborn stuck bullets in point-forming dies is to remove the die from the press, remove the ejection pin, cut a section of oil-quenched spring wire stock of the same diameter as the ejection pin, and simply pound this pin through the bullet from the ejection pin side of the die. It either pushes the bullet out, or it pushes a hole all the way through it. Usually, the lubricated lead core is a better technique. Things to avoid are attempts to drill or gouge the bullet out of the die with a sharp instrument, attempts to heat the die and melt out the bullet (which of course leaves the jacket stuck, a worse problem yet, and also ruins the die temper), or attempts to use acid to dissolve the lead and copper. All these things will result in a spoiled die even in the hands of experts with much experience in removing stuck bullets, in most cases. If the die has to be spoiled, let the factory do it and then it will be replaced by them.

Sometimes a handloader will order only part of a die set, such as a core-seating die or a point-forming die. The straight-walled dies are not as difficult to make as ogive-walled dies; so a good machinist will sometimes try to make his own core seater and only order the point-forming die. It must be understood that there is no way a diemaker can guarantee that the point-forming die will work properly under such circumstances. Swaging dies are made in matched sets. The core swage has to make a core that will fit into the jacket selected. The jacket selected has to fit into the bore of the core seater. And the slug that comes out of the core seater has to be precisely the correct size to slip into the point-forming die, so that the end result is a bullet of specified size. Changing any of the die diameters will change the pressure required to form a given degree of ogive, or tip closure, in the point-forming die. The result is a different diameter and ease of ejection, using the same point-forming die, with differences in the other dies.

About the only guarantee that can be made and backed up for a single die, not ordered as part of a set, is that it will make a certain size bullet when a specified brand of jacket, weight of bullet, and diameter of seated core is used, provided the pressure applied is the same. In other words, before the die can be shipped, the factory has to make up the rest of the die set to test it. If the core seater is not the same as that used by the handloader, then some difference in diameter and ease of ejection will probably result. If by some chance the core seater is too large, then stuck bullets will result and these are not the fault of the point-forming die. The handloader has to take responsibility that whatever seated cores he uses will be smaller than the finished bullet diameter from the particular point-forming die. The difference should be at least a couple of ten thousandths of an inch, and can be as much as a thousandth of an inch.

When troubleshooting die sets to see if they are properly matched, it is important to use the same instrument. The micrometer can be considered a comparison standard between the dies, but the exact reading of the micrometer should not be considered as the final word about the bullet diameter. Far too often, inexperience at reading bullet diameters—which are not just as simple as reading a round rod diameter because of the ogive, pressure ring, and necessary shank taper—will result in errors in interpretation of the readings. A bullet given to several different people will probably result in a range of at least a thousandth of an inch variation, even if all of them use the same accurate micrometer. When dealing in control tolerances of a few ten thousandths, such an error means that the measurements are good only for comparisons and not as absolutes.

Whenever doubt exists about the proper matching of a factory die set, the best thing to do is to make up a core, seated core, and finished bullet, and send all three (preferably several samples of each) back to the factory for interpretation of the problem. This is always a good thing to do, as it allows the factory people to determine exactly which die is at fault and what to do about it. Measurements are almost always taken rather lightly because of the difficulty in getting two people to read the same thing on the same bullet. The actual sample is much better. Comparison to a factory bullet is also a bit difficult with a micrometer, especially since many factory bullets are run through a ring sizer die to put a straight shank on them and eliminate the pressure ring (actually, this is done to control the diameter and remove taper that results from production swaging or hand swaging, and to eliminate some of the requirements for high precisions in the preceding dies). Thus the factory bullet can be miked more nearly as if it were a rod, whereas the hand-swaged bullet should be miked at various points and the readings interpreted from experience regarding desirable taper and pressure ring sizes.

Another common problem with swaging dies is the striking of the bottom punch against the face of the die, so that the punch edge has a burr or nick in it. This doesn't do much harm to a core-seating die, if it isn't excessive, but in a point-forming die the nick is very serious. It causes the base of the bullet to appear to be out of round, and definitely harms the accuracy. The punches should all have sharp edges and be guided gently into the die so that they do not touch the die face at any time. A damaged punch should not be used, but should be returned to the factory and refaced. Punches with cups machined into their faces usually expand at the edges, under the tremendous pressures of swaging, so that they are a very snug fit to the die bore. This is a reason to use the cup base or cup top punch in a die, especially in a core swage die. It helps to automatically adjust the punch to compensate for wear and to seal against the pressure. But removal and insertion of tight-fitted punches can result in nicks to the punch edges. Cup bases that are deep enough can even be bent from insertion at an angle. Therefore, whenever a cupped punch is inserted or removed, great caution should be exercised to do the job in a straight line with the die bore, never using excessive pressure. Quite often the punch will be impossible to remove by hand but will press out only under pressure from the press. These snug fits are best left together. Usually the top punch, which is kept inside the die during all the operations, has the cup. Boattail-forming punches are the exception, and they do indeed receive more damage than similar flat-base dies. Such damaged punches should never be used, but should be replaced with new ones.

Another problem is the use of too long a core or jacket for the die. If a set of dies is made to produce a light bullet, chances are good it will not have the proper punch length and possibly not even the die bore length for longer, heavier bullets. An example might be a .30 Luger bullet swage die used to make 90-grain bullets of .308 caliber. The same die might also produce up to 120-grain carbine bullets, but if a 150-grain or heavier bullet is attempted, the punches would barely enter the die. Applying pressure to them would possibly result in putting a bad scar in the side of them, as they tip instead of moving forward. Also, it is possible to bend the punch if it is too far out of the die when pressure is applied. Swaging dies and punches are made so that a required minimum amount of insertion of the punch into the die results at the maximum recommended bullet weight. Trying to make a heavier bullet can be very bad on the punches.

Another problem with too heavy or even too light a bullet is that the dies are tested over a certain specified range of weights. If a bullet is too light, then the base section may be pressed too far into the die and start into the ogive section of the point-forming die. This would make an undersized bullet. Yet the die might be perfect in every respect. If the bullet is too long, then the bullet may appear oversized because of a slight amount of mouth taper in the die. Some taper near the mouth of the dies is not a bad thing, as it does allow easier insertion and removal of components and helps center the punches. But making bullets that extend into the mouth area because of their length will result in oversized bullets—and probably damaged punches. When checking for bullet taper, one should make certain that the recommended weight of bullet and length of jacket is being used. This also goes for diameter.

A problem which is not really very serious but causes a lot of concern among beginners is the presence of dents or what appear to be scratches on the ogive of the bullet. These are usually running along the bullet axis, rather than around the bullet. There is nothing about a swaging die's manufacturing process that can put a burr or scratch on the side of the ogive. Any factory defects will be in the form of rings caused by chips under the hone or lap edge, or marks from the reamer. Scratches along the length of the bullet are almost always found to be from either the normal folding of the bullet jacket material, from excessive lubrication, or from using a jacket material that is too thick for the amount of closure at the tip. All jacket material will attempt to fold and leave little marks around the ogive. Thinner materials, well backed by lead core, show the least of this. Thick jacket materials, and those with very little lead behind the ogive such as with long jackets and light cores, tend the most toward fold or dent marks.

If one examines a factory bullet, usually the remains of such marks can be seen under the polished shine. The way the factory gets rid of the scratch or fold marks is simply to polish the bullet in a tumbler or vibrator container, using leather scraps, walnut shells, even peening with copper or steel shot. This improves the looks but possibly does harm to accuracy. The bullet swager has to decide for himself if he is after looks or precision. Of course he can have both but the amount of work may be more than necessary for the purpose. The proof of a bullet's quality should not be admiration of one's friends, but the groups on the target or the performance on the game. However, the cosmetics of bullet making can also be done at home, and accuracy is seldom harmed enough to matter. Bullets can be polished to a factory shine in a rock or case tumbler or vibrator. The slight rounding of the base edges will probably not harm accuracy, in most cases.

Another situation is the formation of rings or depressions around the middle of the bullet shank. Since the bullet is pressed in and out again, sliding along the die walls, such depressions cannot possibly be a result of a ring in the die; they would be ironed out in passing the rest of the shank along the same area. But they usually are a result of improper core seating or too soft a jacket anneal. If the core is not seated fully to the diameter of the jacket and then a bit more, to expand the jacket to fit the die walls, portions of the jacket will be unsupported on the inside. When the bullet is pressed into the ogive die, a great deal of resistance is met when the jacket touches the ogive part of the die. At this point, the jacket walls will take most of the force. If they are unsupported by a solid backing of lead core, they will fold or buckle inward. This is one cause of rings or depressions in the bullet shank.

The second cause is the soft jacket. Jackets are work-hardened so that they are stiff enough to take the pressure of forming the ogive. The pressure is applied to the base and is carried primarily by the jacket walls up to the point of contact with the ogive wall of the point-forming die. If the jacket is so soft that this amount of pressure will buckle it, the jacket can fold inward even if backed by a good lead core. This problem occurs more often with rimfire jackets, made at home from fired .22 cases, when they are annealed too much. Actually, only a gentle stress relief is required, not a true anneal. Heating quickly to 400 or 500°F is sufficient. No lengthy process of high-temperature treatment is desired or usable. As long as the jackets will form nice ogives without folding, and yet not take on buckles or wrinkles in the shank during forming, the heat treatment is adequate. This is the final test of the anneal or stress relief. The required time and temperature is very flexible, and is not at all critical, since all the brands of .22 cases are not the same in any case.

Chapter 28

Making Bullet Swage Dies

Of all the products a person might attempt to make at home, bullet swage dies are probably the least desirable, most frustrating, and completely unreasonable project any reasonable person would want for a weekend's diversion. However, bullet swage dies can be made at home. Most of the best dies are, in fact, made in some dedicated and experienced diemaker's garage. The tools necessary for consistently good work are too expensive for hobby use, but even so, there may be some interest in the process, and this is the best place to cover diemaking in a short form.

First, one needs a proper kind of steel. We have found that Timken Graph-Air is the most suitable, since it does not require liquid quenching to harden after heating and thus is not subjected to nearly as much warpage pressure as oil-quenching steels. Also, it contains small pockets of graphite that serve as lubricant and act to trap lubricating films. It can be turned with high-speed tool steel bits and threaded with tool-steel chasers. As shipped, Graph-Air is ready to work.

The steps of making the die are to cut off and face the ends of the

C-H .38 pistol die set.—Photo courtesy of Charles LaPiene, Springfield, Mass.

Clymer pistol dies.—Photo courtesy of Charles LaPiene, Springfield, Mass.

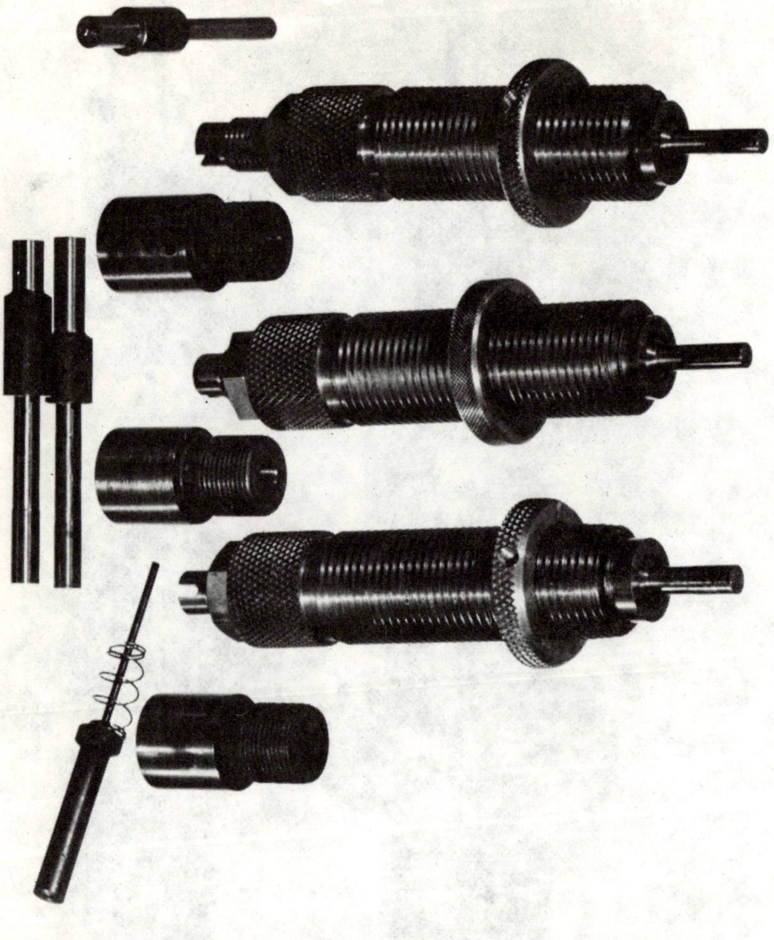

Original SAS. 224 rifle die set.

desired length of round bar stock, of whatever diameter is desired for the press system that will be used, put centers in the ends, and turn the outside between centers to remove the out-of-round condition and the mill scale. The outer skin of this, and most other, tool steels is not the same high-grade metal as the core and must be removed. The

Corbin .224 rifle die set.

outside diameter can vary quite a bit from length to length, unlike most of the cold rolled or ground steel bars, since it is assumed that most people will turn the stock.

After truing up the blank, it is threaded or otherwise fitted with detents for holding it in the die holder, or in the press ram. If it is a straight-walled die, a pilot hole is drilled through the die which leaves about eight-thousandths of metal between it and the finished cavity size desired. Next, two or three reamers are run into the hole to bring it closer to finished size and clear out the drill marks. The first reamer (which is normally a righthand, straight-fluted chucking reamer) cuts two- or three-thousandths away, to clear most of the drill marks. The next is a thousandth larger, and the next a thousandth larger again. These are to remove the reamer marks left by the preceding reamer as much as possible.

To minimize reamer marks, the work should be done either in a very solid vertical mill with a fixture to center the die properly, such as a collet holder, or with collets in a solid toolroom lathe and a floating reamer holder in the tailstock. The speeds at which the least vibration occurs for the particular reamers and machine must be found and used. This holds chatter marks to a minimum. Usually, the reaming is done in very short strokes with the reamers completely removed and brushed off with lubricant, between each pass, while the mill fixture or lathe head is fitted with a pressure-supplied coolant flow. A rotating joint is used at the back of the lathe bore to force coolant up through the pilot hole in the die and carry chips away from the reamer cutting edges. Oil or coolant can be pumped up through most of the better collet fixtures for a mill. A very deep hole, such as an inch, requires a great many insertions and removals of the reamer for cleaning if all chip marks are to be eliminated.

Extremes are taken to eliminate any marks during reaming, since later hand lapping is very tedious and slow. The only way to eliminate such marks is to remove enough metal from the rest of the die to reach the bottom of the mark. Usually, much of a chip mark will go beyond the reasonable limits of metal removal, and result in the die having to be made oversize, which for practical purposes means throwing it away. If one wishes to spend hours lapping the die by hand, one can be less cautious and simply ream the final size five- or ten-thousandths smaller than desired finished cavity size, and lap all the remaining metal out by hand.

After the last reaming, the cavity should be about three-thousandths smaller than finished size. It is next honed using a stone longer than the cavity by at least 1.5 times. The stone is normally an aluminum oxide rougher with a grit of perhaps 280 size. We prefer the use of a Sunnen hand-operated hone so that taper and tight places in the bore can be felt and worked out, although it might be possible to use an automatic if enough metal were removed. At this point, the cavity is brought to within about a thousandth of the finished size. If a honing machine is not available, it might be possible to lap the die cavity very carefully using a hand lap, but taper and out-of-round conditions, or a tight mouth, almost always result if a honing machine is not used. The advantage of the honing machine is that it permits the operator to expand the laps or hones continually within easily controlled limits, as the cavity is enlarged.

The die is next coated with an antiscaling compound such as PBC Compound and put into an electric furnace at 1475°F for a half hour. It is then quenched in still room-temperature air simply by being left alone, supported by a wire or on a mesh that allows the air to circulate around it. After it cools to room temperature, it is put back into the furnace and drawn at 500°F. After a half hour at this temperature, it is ready to finish. The finish hone is silicon carbide, of 400 grit. The die is honed to remove all the metal left between the desired size and the size after heat treatment, to remove any remaining taper in the hole, and to correct any out-of-round condition that may still exist.

At this point, if any flaw requires removal of more than about a half a thousandth of an inch of metal from any one point on the die walls, the die is ruined. This is because it is within a thousandth of bore size at this point, and it cannot be used if it is oversize. Depending on the jacket and the use, it may be desired to polish the die with either a very fine diamond or crocus to remove not more than the last ten thousandth of an inch. Usually this would not be necessary in a core-seating die. The antiscale compound is removed by boiling the die in water for several minutes. Whenever the die is left for any period of time, the bore should be oiled to prevent rusting.

The die can only be tested after punches are made to fit it, and a press and system of holding it are devised. Punches should be a snug but sliding fit in most cases. Generally it is safer to draw them to a lesser hardness than the die body, so that they will not be as likely to damage the die as to be damaged themselves in the event of a mishap. Punches are much easier to replace. The straight-walled die is the easiest die to make and can be made by any dedicated hobbyist who will take some pains not to get chips in the reamers or allow the honing or lapping to get the die out of round or tapered through improper procedure. The point-forming or ogive-walled die is the one which is the most difficult.

There are several problems with the making of point-forming dies

Straight-walled die cut-away.

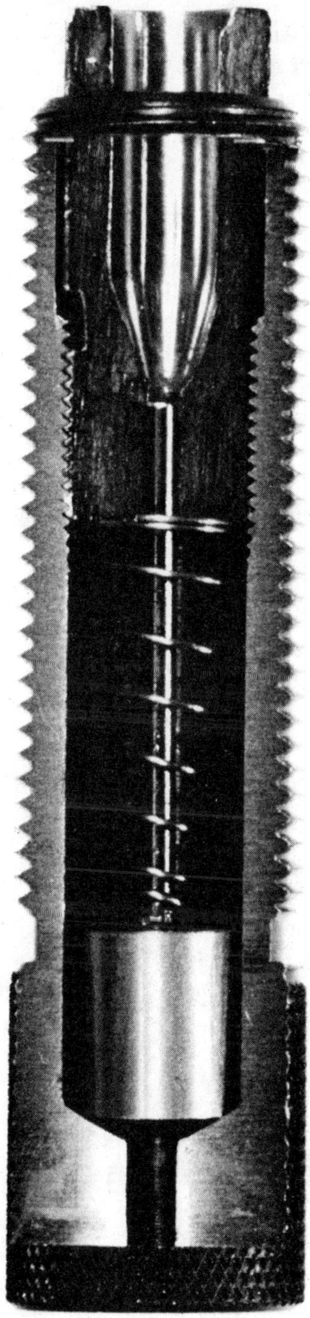

Ogive wall die cut-away.

which generally preclude these being made at home by the hobbyist. The cavity is blanked out in several steps using a series of drills, and the smallest one is the size of the ejection pin. This usually means a deep drilling with a very small number drill, in tough tool steel. The problem here is broken bits stuck in the hole, which can be eliminated only by keeping the bits sharp, throwing them away after only light use, and by very frequent cleaning of the bit during the drilling process. If any of the larger drills is run too far into the cavity, it will leave a serious ring in the finished die wall, which ruins the die. Careful calculation of depth is necessary if a person is to use roughing techniques to the fullest advantage, which is to keep simply the amount of metal removed by reamers at a minimum.

The second large problem involves the procurement of a properly shaped set of reamers. There is no store or supplier which will offer a 6 caliber ogive spitzer reamer in exactly 0.2230 inch followed by an identical finish reamer at 0.2235 inch, or any such combination of tolerances and shapes. Commercial multifluted reamers are expensive because they are each custom-made to specifications, involving a considerable delay and a need to know how to specify the various parameters of a reamer. Almost no supplier will guarantee that the reamer will cut to within a half-thousandth or even a thousandth of the specified size, though he may guarantee the reamer is itself a certain size.

One of the most frustrating problems with precision drilling and reaming to exact sizes is that the size of the reamer or bit is usually different from the hole it makes. When dealing in fractions of an inch like a 1/2-inch hole, a drill that is half an inch in diameter cuts close enough; but when the diameter is supposed to be precisely specified to the nearest thousandth or half-thousandth of an inch, it is a matter of trial and error. Slight vibration, off-center point, one lip slightly longer than another, one edge slightly duller, and the tool bit will make a larger hole than it should. Even the speed of cutting and the amount of coolant used will cause a change in the hole size. The answer is to make one's own D-reamers, which are tool steel bits shaped like exactly half of the bullet, cut precisely down the center and given a slight relief for the cutting edge.

This requires either a very good tool and cutter grinder, optical comparator, and the skill to use them, or a little luck and a very good hand and eye. One can turn the reamer blank on a lathe, shaping it like the bullet, and polish it to the desired diameter of the cavity less about a thousandth for overcut. Then the tool steel can be hardened, and later ground in half, or else milled in half while soft and hardened in this shape, which does tend toward warpage of the tool. A good antiscale compound is a necessity with these highly finished parts. The final surface should be polished with at least 600 grit (crocus). If all goes well the reamer will cut properly when the edge is honed and relieved. This is done by hand under a good light, except where the necessary equipment is available to do it using mathematics and setups. Nothing more than a hand grinder might be required in the first instance, while a tool and cutter grinder costing as much as most automobiles are worth would be required in the second.

The roughed-out hole in the die is reamed to produce the cavity shape and size, less at least a thousandth for hand lapping. If the reamer does not make marks over a half-thousandth deep, the die will be ready for diamond lapping after it is hardened. Unfortunately, one does not know if the die is good or not, for having such marks deeper than can be lapped out within the amount of surplus metal left, until the die is ready to make bullets. This means that days of hard work making reamers, the die, lapping it using handmade laps, and making the punches that fit it can be wasted when the bullet simply fails to eject or form properly in the perfect-looking die. To avoid this as much as possible, it is necessary to force coolant up through the pilot hole in the die during reaming, and to take only very small cuts, removing the reamer to clean the edges after each few thousandths of an inch.

The reamed die is hardened, and annealed or drawn as previously described. A series of brass laps are made using exactly the same ogive as the reamer. This means a way to duplicate the precise curve each time has to be devised, since doing it by eye is usually not adequate. A jig that fits the cross slide of a lathe, and swings a tool bit around on a pivoted arm so it can scribe any portion of a circle's arc at the surface of the stock, is one good answer. The arm length and position can be made so that the same ogive is cut time after time. The laps are exactly like the reamer in shape, except that they are slit down the center and have a small hole drilled through them lengthwise all the way.

There are several ways to use hand laps. The best way is to put the lap in a device that will rotate it and allow the operator to expand it while it is turning. A honing machine is designed to do this with honing stones and mandrels. An adapter can be made that will allow a brass lap to be expanded in the same manner, using a tapered pin in the hole through the lap. This same basic lap design can also be used for noncontinuously adjustable lapping devices, but it requires a larger number of laps or a great deal of wasted time expanding one lap as it wears. Depending on the surface finish necessary, one may use diamond abrasive compound from 40- to as little as 3-micron particle size. A micron is only 0.001 millimeter. Many germs and viruses are larger than the smallest diamond particles used.

Diamond compound is put on the lap by pressing it against a hardened steel surface with a small amount of the compound spread thinly on the surface. This is a somewhat time-consuming procedure if done correctly, as the diamond is not rolled or painted on, but pressed straight down into all the surface of the lap with firm pressure. The diamond will imbed itself into the softer metal, and some particles will be held so that their edges are above the surface, making tiny cutters. Once a lap is loaded with diamond of a certain size, it is marked and kept for that size only. A normal progression might be micron sizes of 40, 30, 20, 15, 10, 5, and 3, although one can skip from 40 to 20, and from 20 to 10 microns by spending more time on the smaller size. Skipping too large a range will not create a true smooth surface.

It is very easy to make metal shine brightly, simply by polishing most of the scratches from the surface and polishing with crocus cloth. This is not the same thing as a true smooth surface such as is required for a good-quality, smooth-ejecting bullet swage die. A surface can be full of waves and tiny ripples yet appear very shiny as long as there are no sharp edges to break up the light and spoil or diffuse the light reflection. Such a surface can be made in a die cavity with various grades of abrasive paper used on a stick, but the die will have any or all of these faults as a result: wavy surface, out-of-round shank, ogive not concentric with shank, ogive shape different from the one intended, excessive taper, possible reverse taper, ogive larger than shank, tip of ogive not polished, tip polished but start of ogive rough, or oversize and undersize sections.

The correct size of abrasive to polish a surface is the size that is just smaller than the roughest marks are deep, so that the abrasive will not just ride the waves, but will actually be large enough to cut into and remove them. When this size of scratches is removed, the size of scratches left will be that of the abrasive particle size. Then another, smaller size is used to remove these. By working to smaller and smaller sizes, eventually the scratches are made so much smaller than they were originally, that the surface appears perfectly smooth to the eye. But even the smallest abrasive still acts by making scratches and removing chips the approximate size of the abrasive particles. No polishing act will add metal to fill the grooves. The act of burnishing is a method of approaching this.

Burnishing is using a dull hard tool or material, such as a special abrasive, to push the higher ridges of the metal over into the lower areas. It works only after the metal is polished to such a fine degree that only a microscope could show the burnishing effect. Generally speaking, a small amount of automatic burnishing can take place when the finest grade of diamond is used, since it becomes clogged with chips so easily (these chips being smaller than dust, of course). We have not found it necessary to resort to special abrasives after the finest diamond grade is used. Generally, the final grades are used on materials softer than brass, since the size of the chip is determined by how much of the diamond particle is above surface. Softer materials, like plastic, wood, felt, or cotton, allow less and less support of the diamond so that it is pushed back away from the surface. This means that one grade used on different lap materials will cut different surfaces and give different polishes.

Because it is necessary to control the taper and shape very closely when using a lap with fast-cutting, larger grits of diamond, only the finest grades are used on nonexpanding solid laps. An expanding wood or felt lap would be hard to make and control in any case; so the grades which cut a very small amount compared to the shape and diameter of the die cavity are used on these laps. This prevents changing the cavity dimensions by more than a minuscule amount. Pushing a wood lap into a die cavity, having loaded the lap with 40-micron diamond, would enlarge the mouth first and then spoil the ogive by wearing away at the junction of the shank and ogive. But doing the same thing with a 3-micron lap would merely polish the mouth slightly more than the rest of the die and not enlarge it enough to damage the die. A properly fitted wood lap would polish most of the die evenly and an improperly made one would not do significant harm.

The problem with lapping is to remove enough metal so that the surface finish approaches final polish just as the diameter approaches the desired size. Special care has to be taken so that the mouth is not lapped more than the rest of the shank, so that taper does not become excessive. Generally, it is best to bring the die to within a thousandth of an inch of finished size (or with questionable reamers, two-thousandths) and then lap the remainder out. About ten to fifteen minutes of lapping with the larger grits will remove a thousandth of an inch of hardened tool steel. This means continual checking and testing of the die while the final lapping is underway. While it would seem that a plug gauge or some other simple means of measuring the cavity would be adequate, there are too many factors that affect the bullet size besides the cavity size to rely on this.

First, the size of the cavity will usually be larger than the bullet it makes because of jacket springback. This will change from brand to brand of jacket. It will also change from one lot to another with some brands. The amount of lubricant and the pressure it requires to form the desired amount of ogive will determine how much pressure is applied to the shank. This in turn determines how much larger than the seated core and jacket the shank will become in the ogive die. A very smooth die might make a smaller bullet (in the shank area) than the same core and jacket used in a slightly rougher die. This is because the pressure is more for pushing a jacket and core into a rougher die, expanding the shank more. So, while lapping the die with progressively smaller grades of diamond, one is faced with the problem of how to know when to quit. The very fact that there is diamond in the die, stuck in the die walls, will make a die eject and form bullets slightly harder, requiring more pressure, and thus making a little larger bullet, than when the diamond finally is pulled out by the passing of many jackets through the die.

Of course, before the point-forming die can be finished, it is necessary to have the core-seating and core-swaging dies done, punches made and fitted, and a supply of seated cores made up for testing. This means a quantity of proper jackets has to be on hand. An attempt to make a point-forming die without all these things, plus the punches for the point-forming die itself, means any success is purely a matter of luck. The same cavity size will make a larger bullet and also eject harder when the core-seating die is nearly the same size as the cavity in the point-forming die, than it will when the core-seating die has a cavity perhaps a half-thousandth under point-forming die size. This is one reason a single die, not supplied as part of a set, can't be guaranteed to make a certain size bullet or even to work with some unknown set of dies. To be assured the dies will eject properly and make the correct diameter of bullet, the core-seating and point-forming dies must be matched very carefully.

In bringing the die to final size, the smaller grits or grades of diamond cut slower and slower. This gives the diemaker more time to fit and test, but it also means that unless one has nothing but time to spare, the die should be within a half-thousandth or less of final size (determined only by continually cleaning out the die and making some bullets and then measuring the bullets) when the final grade of diamond is being used. Otherwise, what should be a half hour to an hour's job becomes an all-day operation. Also, the die cannot be tested until it is at least smooth enough to form and eject bullets. This starts to occur after the 20- or 15-micron diamond is through. If by then there is excess taper or the diameter is too close on the bullet being made, chances are that the die will be oversize or out of tolerance by the time the final lap is through. Since a hardened steel die can't be reamed without spoiling the reamers and throwing up more chips and gouges than is acceptable, and since it is so difficult to soften air-hardening steel, the only answer is to throw away the die, along with all the work that went into it.

Quite often the question is asked, why not enlarge a point-forming die with a flaw to the next caliber, like rechambering a rifle? If it were easy to soften the kind of steel used, and also if it were practical to try to guide a reamer down the type of ogive already in the die (sometimes it is, and sometimes not) without leaving flaws in the new surface, or removing too much metal for the length of the die in order to clear the old chamber or cavity, then this could be done. Generally the cost of the work is so high compared to just blanking out a fresh die and starting with it, that there is no particular advantage. With air-hardening steel, it is not practical since the time and electricity would be excessive to soften the die.

Bullet-swaging dies can be made by other techniques, including exotic electrochemical and electrical discharge machining, but there is no easy answer to maintaining the tolerances, surface finish, taper, and concentricity. Jig grinders and other expensive machinery provide alternative diemaking methods. In the final analysis, hand lapping with diamond is the final step, while trial and error is the final test. The proof of the die is always the bullet it makes, never the cavity itself. Experience with dies can give a diemaker a good idea of the probable performance of a die based on the cavity measurements. But only making a bullet will actually test the die.

If it sounds as if diemaking is not being presented in a very enthusiastic manner, or as if there is intent to discourage experimenters, it is only because, having spent so many years and so much money attempting to overcome seemingly insignificant problems involved in making dies, we do not want to be guilty of causing anyone to feel that diemaking is a good weekend hobby. Chances are that the first die, and the second, and probably the hundredth, will entail as much lost time, broken tools, and frustration as any hobby one could choose. However, we do have die blanks and standard parts for those who wish to pursue it. It is a wise idea to try to build a die on a standard plan, using standard parts as much as possible, since these have proven satisfactory in years of operation, and are generally made available cheaply compared to making them at home.

Periodically requests are received for plans, parts, or advice on making dies that would fit a hydraulic press. It seems many handloaders regard a hydraulic press (which is not difficult to rig up using air-over-oil cylinders or a simple truck jack) as akin to the magnum calibers—a source of extreme power and pride, worthy of a bit of bragging. There is no particular advantage to a hydraulic press over a well-designed mechanical press, for swaging small arms bullets. The hydro-press is usually slower, and much more dangerous to the user and to the punches and dies; any attempt to apply more pressure than is required to bring the core and jacket to the size of the die bore is useless in any case. In point-forming the use of extra pressure only results in putting a little pipe on the bullet tip (by forcing jacket material up the ejection hole). Hobbyists are best advised to use mechanical presses; swaging isn't hard and at least a few of us could use the exercise.

Chapter 29

The Lead-Tip Die

Although lead-tip bullets can be made in almost any die set, the ejection pin in the point-forming die limits the shape and size of the tip. Since the pin presses on the lead tip to eject the bullet, a few point-forming dies may not make good lead-tip bullets because they require more ejection pressure than the lead can stand. A depression in the lead, or perhaps even a pin stuck into the bullet tip, may result. To form good sharp or round ends on the lead tip, a die called a lead-tip-forming die may be used.

This die is used after the point is formed. The core and jacket length ratios are carefully adjusted so that the correct amount of lead is left extending beyond the jacket. This extra lead is shaped in the tip-forming die to the desired shape. The bore of the tip-forming die is a straight-walled hole, slightly larger than the finished size of the bullet. Into this cavity is fitted a top punch with the shape of the bullet tip reamed and lapped into it. The tip-forming die operates very much like a core-seating die used to make handgun bullets. The finished bullet is pushed gently up into the die cavity, where the extra lead at the tip is formed by the cavity in the top punch.

If too much pressure is applied, two things will happen. First, the edges of the punch will contact the jacket and put a depression in the ogive. Second, the bullet base will be swaged up to the size of the die cavity, which will make an oversized bullet. The punch used to push the bullet into the tip-forming die can be the same one used with the point-forming die. The punch only has to push the bullet very gently into the die and bump the soft lead nose into the cavity within the top punch.

There is another way to finish the lead-tip bullet, requiring no extra die set. If the bullet is seated in a case with a normal or tight neck, the seating punch will have enough resistance against it to impress a new form in the soft lead tip. By making the bullet-seating punch cavity the correct form, the bullet can be finished in the process of seating. Nice spitzer tips with sharp or slightly rounded ends can be made in this way. The same ogive is continued into the punch cavity as is formed by the jacket of the bullet.

Generally, the same reamers are used to cut the point-forming die cavity as are later used to make the tip-forming die. This assures that the shape and size will be correct. If the shape is not the same outline as the original bullet ogive, the lead tip will take on a separate ogive of its own, which will be an abrupt change from the jacket shape, and will look strange. Generally, the depth of the cavity is from a sixteenth- to a quarter-inch at most. This very small cavity is difficult to shape precisely except by hand labor, even when roughed out with the same reamer as was used to make the point-forming die. Generally it is best to have the point-forming die and lead-tip-forming die made and tested together.

Since the lead tip is so easily formed by light pressure, some shooters wonder why it would not be possible to make a new top punch for the core-seating die, and use this die as a guide to press the bullet into the punch cavity. The reason is that the core-seating die is smaller than the bore of the point-forming die. The finished bullet would never slip back into this die, and the pressure required to get it in and to eject it again would result in a ring being pressed into the ogive where the punch ends appear. The bullet would also be undersized even if it could be ejected without damage.

One form of lead-tip bullet is actually easier to make in the regular die set. This is the flat-nose bullet, either with or without a hollow point. To make a good flat-nose bullet, the lead is seated slightly longer than the jacket, and the ogive is formed partially. Then the bullet is ejected, turned end for end, and put back into the point-forming die. It is bumped very gently with the bottom punch against the lead tip. The die merely holds the bullet so it is straight. The bottom punch (which normally presses on the base) makes a good flat-tip bullet, requiring no die changes or extra parts. Bullets of the .30-30 style are especially good for this process.

When lead-tip bullets are going to be made exclusively, a special point-forming die with a larger than normal ejection pin, perhaps larger than 0.1 inch diameter, can be supplied. This makes it easier to eject without penetrating the lead tip, and gives a larger flat area and a larger ejection pin hole into which to extrude surplus lead. Usually the round-nose styles use a larger ejection pin than the spitzers for this reason. Round-nose bullets are more often made with large lead tips. The lead-tip-forming die is an option which is required only in a small percentage of cases, and is actually useful more to the commercial bullet maker, whose bullets must look finished before they are loaded. In practical effect on accuracy or performance, the lead-tip die has little to offer over the normal three-die set which can make lead-tip bullets having a small flat on the end. Where factory-finished-looking bullets are needed, this die will do the job well.

Chapter 30

Bullet Swaging For Profit

An interesting aspect of bullet swaging is making custom-made bullets for resale. Under current laws, an ammunition manufacturing license is required by the Bureau of Alcohol, Tobacco, and Firearms (BATF) to engage in making and selling bullets. No license is required to make bullets for one's own use or for the use of a private club. The tools and components for swaging, including bullet jackets, are universally shipped and sold all over the United States without restriction. At this time, the license required to sell bullets is inexpensive and no more difficult to obtain than a regular dealer's license.

The key to success in selling custom bullets is that they must truly be something different, not just a cheaper duplicate of a factory bullet. Making bullets by hand is only economical compared to purchasing the more expensive kinds of factory bullets. Ordinary hunting grade bullets can be made much more cheaply by the factories and there is no real way to compete on price with homemade bullets. There are a goodly number of bullets available in nearly all calibers from the factories, but there are also a fair number of special calibers and oddities that are just not quite popular enough for mass production and marketing. These special calibers are ripe for the home swaging market.

To sell bullets one manufactures at home, after obtaining the proper license, one need only make sure of his quality and consistency by doing thorough tests of his bullets, and then purchase an ad in any of the better national magazines or regional sporting papers. Bullets can be sent to any dealer licensed to sell guns and ammunition; so a dealer discount is a virtual necessity for all mail order sales. The packaging and promotional materials can be as simple as a poly bag and a tagboard insert, or as fancy as the new business can afford. The success of the venture as a part time money-maker can also be based on local sales, where conditions of hunting or target shooting present a need for a special bullet design, even in a standard caliber.

Some ideas of bullet calibers, and special designs, that have been successful in various areas include the .20 caliber, the .15 and .17 caliber, and other special subcaliber bullets. Usually these are offered in weights that are not available from any factory, and if the bullet maker teams up with a gunsmith who can do conversions or build these special rifles (or handguns), a continual source of mutual support for the idea results. Custom bullets are the key to success for some custom rifles, such as the .224 Clark. These require bullets which are much longer and heavier than any standard factory bullet, using a fast-twist barrel.

Other ideas include making very highly explosive bullets out of rimfire jackets, or bullets using multiple jackets for very effective long-range game hunting, or other special designs to fill local needs. For some time it was popular to purchase a certain brand of wadcutter bullet which had a hollow base, loading it backward for police use. How much easier, cheaper, and more effective it would have been to use custom-swaged bullets with the cavity made in the nose, exactly as the local police desired, and in any caliber. Many articles have appeared about the popular use of the new light frame .44 Special revolver as a home defense weapon, but at the same time noting there is no effective bullet that will expand at these speeds. A deep cup with lots of exposed lead in a jacketed .44 bullet can be swaged at home, and it will expand almost upon dropping to the floor! A custom bullet swager can take advantage of the lack of effective bullets in many calibers.

One of the biggest mistakes a custom bullet maker can make is to price his bullets so as to compete with standard factory products, because he cannot compete with a bank of punch presses cranking out bullets which are designed to sell to a large national market. By offering bullets to fill special needs, which are otherwise unavilable at any price, a person can actually make a profitable income from bullet manufacturing on a small scale. But one of the best things about bullet making is that it can be such a good way to use time that otherwise might go to waste, especially the time of hired help in a small shop. Often a gunshop consists of one gunsmith and enough work for one and a half people. That is, there is almost enough income to pay for an extra worker, but because it is hard to find a helper dedicated and knowledgeable enough to carry on the same kind of work, only the less technical tasks can be done by hired help.

This means that the gunsmith is busy all the time, and the help only busy during the rush seasons or half busy waiting on customers and doing minor chores. By adding a set of three presses to the bench, and spending a few hours training, the gunsmith can give his help a profitable way to spend the rest of his time, build a name for his shop as a maker of custom bullets as well as a gunshop, and thus afford full-time help even for half-time chores. Also, the family business can become a little more profitable and draw the members a bit closer when bullet manufacturing is added to the activities list. Anyone, even a child, can make bullets once the presses are set up correctly and some time is spent showing what is wanted and how to get it. Husband and wife, son and father, even father and daughter businesses that thrive on summertime work making bullets, and then selling the bullets in the fall and winter, are not uncommon today.

On the local level, word-of-mouth advertising may be enough. If this is the case, the total cost for starting a bullet-making business, or adding it as a profitable activity, can be very low. One can start with a single press and set of dies, some lubricant, a supply of bullet jackets and lead wire plus a core cutter, and some boxes or bags for the bullets. Usually one can get by for less than $300 at this time (which, with today's inflation rate, may soon be a higher figure; assume a week's wages for a skilled machinist as a guide to the minimum cost).

There are very few activities that promise any sort of return which can be started for less investment. What kind of return can be expected? Since the labor is the biggest factor, this will determine the return. Lead typically costs about a cent per bullet on the average, which is taking everything from the big .44's to the little .17 caliber bullets and striking an average. Jackets usually average between two and five cents, depending on the size and caliber. This makes the material cost between three and six cents per bullet, again speaking in very general terms. Some lighter bullets cost much less; some big ones cost more. They typically cost half of the price of the nearest comparable factory bullet.

The key then is how fast a person can make bullets. One person can easily make a hundred bullets in an hour, each bullet requiring three dies. But with three presses, three people can make more than three hundred bullets an hour because the setup and adjustment is only done once, and a person can become very efficient at handling one step rather than handling the material so much between steps. In fact, one person and three presses can turn out at least two hundred bullets an hour by himself. This is an average, again, and it can be more or less depending on the particular bullet. We are speaking of a hand press made for swaging, with automatic ejection, like the Mity Mite System. One should really not consider a reloading press for commercial work, as it is too slow in the long run.

Typically, custom bullets can be sold for an average price of between five cents and fifteen cents each, depending on the design and purpose. If one person can make two hundred bullets an hour, which cost between three and six cents each, and then resells them at the indicated prices, this means that the hour will yield between $4 and $18 net profit before packaging is taken out. Priced low, the operation will only make spare cash for the business that is better than wasting time between customers; but priced right, the bullets can bring in a good income that helps pay the overhead for the rest of the operation.

How can one sell bullets at prices higher than factory bullets to make a good return on the time and investment? First, by offering calibers and styles that simply are not available from the factory. Second, by offering custom weights and nose shapes, such as extrabig hollow-point pistol bullets, or very light automatic pistol bullets, to the customer's own specifications. Third, by demonstrating performance that is far and beyond what might be expected from a factory bullet for a given purpose. The latter is not hard to do. Multiple jackets make far better partitioned bullets than one can buy; yet they are very easy to make. Many calibers such as the .25 ACP simply have no suitable bullets for defense, period, and yet fired shotshell primers can be made into excellent jackets that will open up.

To show one's wares, one must be a shooter and be able to demonstrate to friends and customers, to show fired bullets and cut-away examples in the shop, and to convince a knowledgeable shooter that the design is better than anything else on the market. This is hard to do unless it really is a good design, and, this means not everyone's ideas will sell. But those shooters who actually recognize a need and then work toward filling it, with a bullet that actually outperforms anything else available, can name any reasonable price and get it. For daily practice and plinking, nearly all shooters want the cheapest price possible, but for matches and big game hunts, there really isn't a bullet too costly as long as it performs well.

Bullet swaging can bring in another kind of profit for the gun shop. Even if the actual bullet sales themselves are not great or particularly profitable, the reputation and interest which will result from having a custom bullet maker located nearby will bring shooters to see the operation. And bringing shooters to the store is half the battle in selling other things. Having swaging equipment "up front" where the operation can be seen also means that the gun shop can put on demonstrations any time, bringing special crowds for the event and coordinating with sales of other products. It is so unique and different that swaging seminars nearly always bring out the dedicated handloaders in droves, and it is precisely these people who are the best gun and component customers.

One of the best show stoppers at a gun show has proven to be a continual swaging demonstration, especially one in which bullets are made from fired rimfire cases. A lot of nonshooters and nonreloaders stop and wonder at the process, but the handloaders in the crowd almost always flock to the demonstration, tell their friends, and bring in more handloaders the next day. A gun show is a good place to make initial contacts with local customers, and to show off the process as well as the products.

Chapter 31

The Accuracy Of Mutilated Bullets

The following report is by John Nosler, courtesy of Nosler Bullets, Bend, Oregon. It points out the important effects which bullet-tip damage can have on accuracy, something which has largely been overlooked until now. Bullet swagers will find the information and tests useful in following the lead Nosler has established with its soft-point designs.

Certainly one of the most important, if not the most fascinating, elements of bullet making is the performance testing. At Nosler Bullets, Inc., we routinely conduct thousands of closely controlled performance tests, many of which are designed to show the effects of surface imperfections on the bullet's flight.

Mutilated bullets are actually more common than most shooters realize. The noses of soft-point spitzer designs can become dented or battered from working through the action; in addition, magnum rifles may tend to flatten the points of those rounds remaining in the magazine, especially if there is no provision for point protection. Testing these common forms of mutilations ultimately led Nosler to introduce a line of .30 caliber Partition bullets designed to withstand most types of point damage. These "Protected Point" bullets were intended for use in the short-necked magnums (where overall cartridge length can be so important) as well as in those rifles where point damage is a common problem. The durable "Protected Point" design

The Nosler "Protected Point" Partition bullets are built to withstand the abuse of rugged hunting trips and still yield precise accuracy and dependable expansion.—Photo courtesy of Nosler Bullet Company

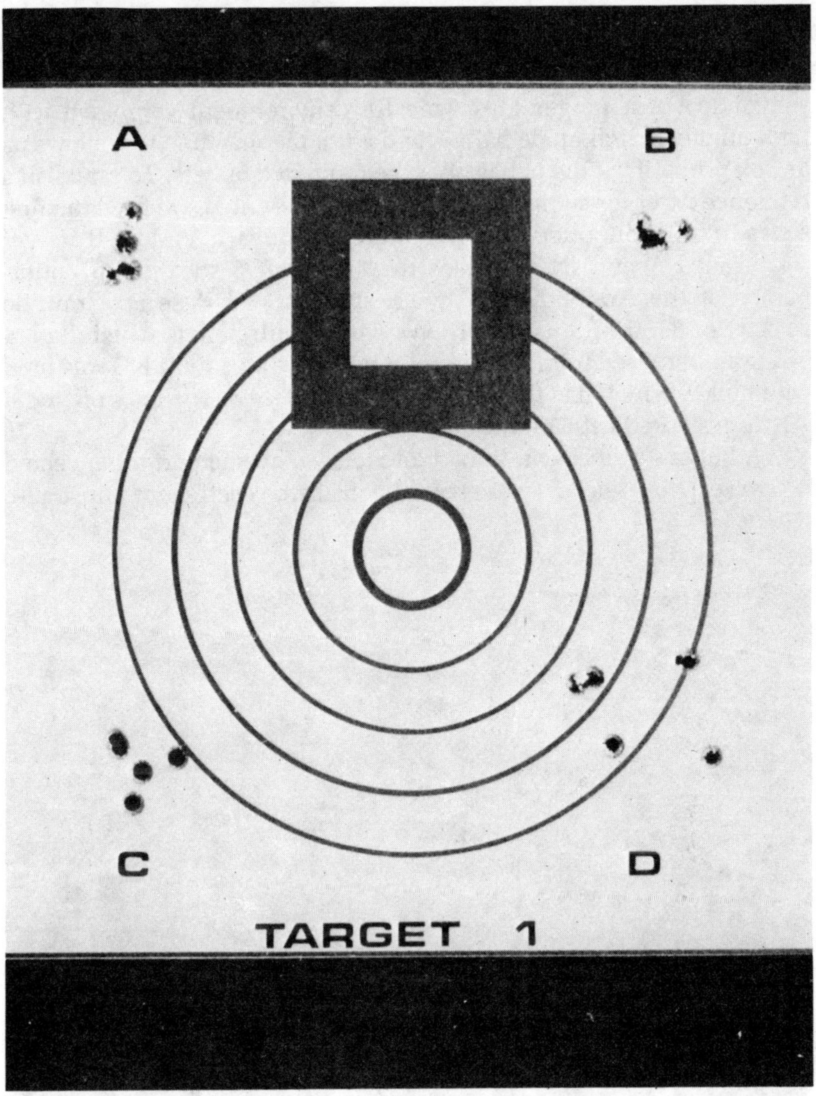

Test groups 1-C and 1-D show the effects of basic point mutilations. Groups 1-A and 1-B represent the normal performance of the test bullet.—Photo courtesy of Nosler Bullet Company

will withstand the abuse of many hunting trips without yielding to damaged points; also, since the point was designed and manufactured to this configuration rather than battered to a similar shape, its consistent accuracy in the field can be a very important feature to the serious hunter.

The tests included in this chapter were fired at the Nosler 100 yard indoor test range. This wind free environment combined with high quality match grade barrels and a unique machine rest allows us to look primarily at the behavior of the mutilated bullets. To establish a reference point for comparison of the tests presented, we fired the first series in a match grade .222 Remington. Groups 1-A and 1-B (target #1, groups A and B) represent the "Normal" 5-shot group performance of the Nosler .224-50 grain spitzer Solid Base at a muzzle velocity of 3100 FPS. Group 1-D was fired with bullets which had .4 of a grain removed from one side of the nose while group 1-C was fired with bullets which had the noses severely flattened. The tests on target #1 were all fired with identical loads.

A bullet's shape is an important factor in overall performance and whenever that shape is altered, the ballistic coefficient must also

The bullets used in test targets #1 and #2. Left to right: *point design of a typical Nosler .224-50 grain spitzer Solid Base bullet as compared to the same bullets with a flattened nose, and with one side of the nose removed.*—Photo courtesy of Nosler Bullet Company

change. Basically, ballistic coefficient is a value used to indicate a projectile's ability to overcome air resistance in flight. Therefore, when the points of the .224-50 grain bullets were flattened, the ability to overcome air resistance was significantly reduced. The normal ballistic coefficient for the Nosler .224-50 grain spitzer solid base is .244; yet, the coefficient of the flattened test bullets used for group 1-C (target #1) was .177. The most dramatic effect of such a change in ballistic coefficient would appear if the shooter was sighted in for the undamaged bullet (.244 coefficient) and fired the bullet with the .177 coefficient. Aside from the change in impact which would result from the bullet's somewhat more erratic behavior, as shown in group 1-C, the drop from the bore at 300 yards has changed from 21.9 inches to 25.2 inches—an increase in drop of 3.3 inches in the .222 Remington. Table I shows the potential effects of mixing normal and mutilated bullets in the same group.

TABLE I

Bullet	*Ballistic Coefficient*	*Muzzle Velocity*	*Range*	*Remaining Velocity*	*Time of Flight*	*Drop From Bore*
Nosler .224-50 Sp.	.244	3100 FPS	300 yds	2000 FPS	.3623 sec.	21.9″
Nosler .244-50 Sp. w/severely flattened tip	.177	3100 FPS	300 yds.	1667 FPS	.3983 sec.	25.2″

Group 1-D, where metal was actually removed from one side of the nose, is the result of a severe disruption to the axis of rotation. In contrast to this, group 1-C, where the metal in the nose was pushed back in a fairly uniform manner, showed a definite tendency to group tighter.

Since the accuracy of point mutilations appears to be a function of its effect on the axis of rotation, a higher velocity combined with the same twist rate should cause the bullets to disperse even more. To prove this theory the entire test series was conducted again (target #2). This time we used a .22-250 to increase the velocity without changing the 1-14" twist rate. By increasing the velocity, we have actually increased the rotational spin imparted to the bullet. This will tend to magnify the out of balance effect in the bullet and enlarge the group size.

One of the most common point mutilations is a slightly bent nose tip. This condition is achieved by carelessly running a cartridge

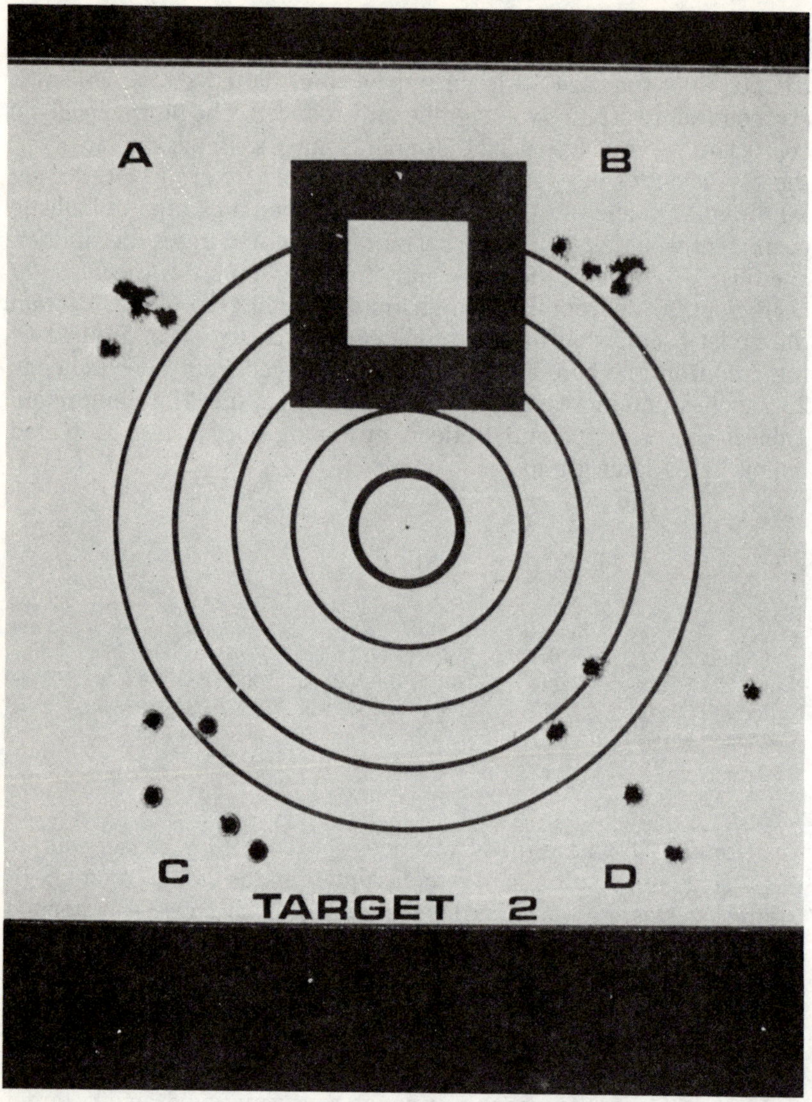

To prove the effects of increased rotational spin, target #1 was refired using a .22-250 at 3700 FPS.—Photo courtesy of Nosler Bullet Company

through the chamber or by "jamming" the bullet against the bottom of the seating die during the reloading operation. The control bullets for this test are the new Nosler .30 caliber 165 grain spitzer Solid Base, a very accurate bullet designed especially for shooters who demand high performance at an economical price. Groups 3-A and 3-B (target #3) show this bullet's normal 5 shot accuracy at 2750 FPS muzzle velocity. The rifle was a .308 Winchester, a caliber that should allow us to draw some valid conclusions due to its long history of accuracy. As seen in the photograph, the bent nose test (group 3-C) almost turned in

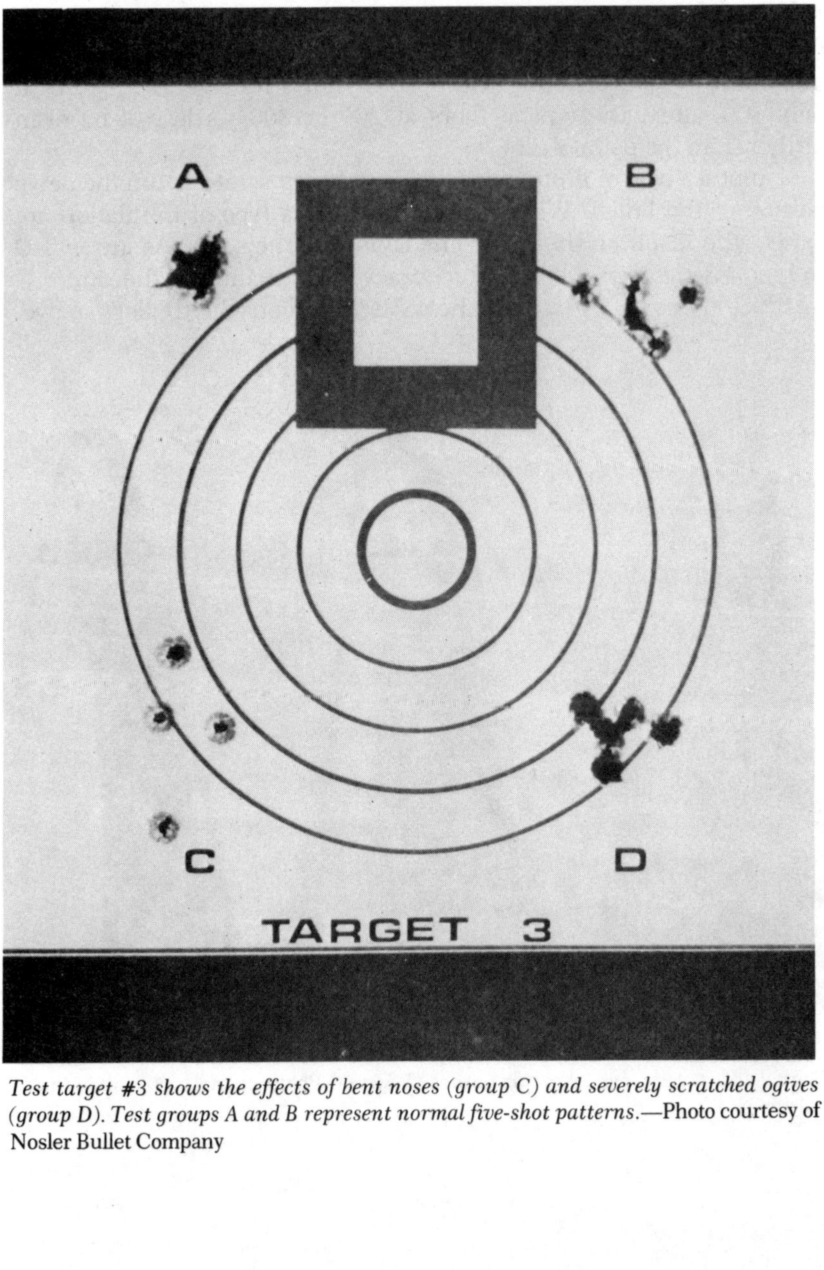

Test target #3 shows the effects of bent noses (group C) and severely scratched ogives (group D). Test groups A and B represent normal five-shot patterns.—Photo courtesy of Nosler Bullet Company

passable accuracy; however, after repeating the test several times, it became apparent that the normal grouping capability of the bullet was severely altered and that the usual consistency was always disrupted by at least one flier. Bullets with this type of mutilation seem to be more common among cartridges that have been around the house for a while, haphazard handloads, or occasionally in factory ammunition. From test group 3-C, the only practical conclusion might be to reject those bullets with bent noses unless predictable accuracy is *not* important to your shooting. Group 3-C shows the average results at only 100 yards; the typical fliers at 200 or 300 yards will be even farther from the point of aim.

Another type of mutilation is folds or heavy scratches on the ogive portion of the bullet. While the causes of this type of mutilation are varied, the result on the firing line is usually the same. As group 3-D (target #3) shows, the usual accuracy performance of the bullet is unaffected—even by a very heavy scratch on the bullet's ogive.

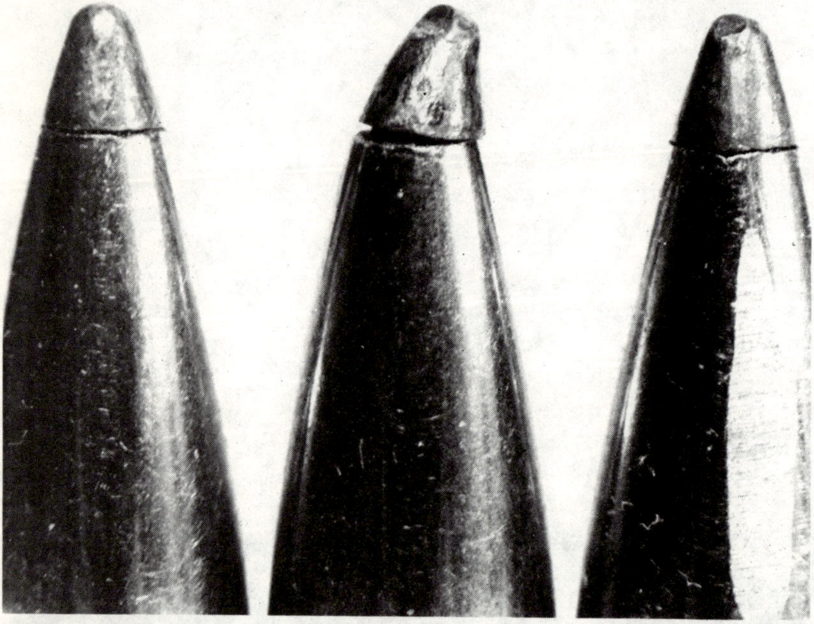

The Nosler .30 caliber 165 grain Solid Base bullets used for test target #3. From left: *normal bullet, bent nose, heavily scratched ogive.*—Photo courtesy of Nosler Bullet Company

Although scratches or marks of this type did not significantly affect the performance of the Nosler .30 caliber 165 grain Solid Base bullet, surface imperfections of this magnitude can be very detrimental to accuracy when found on bullets where the natural stability is not sufficient to overcome such flaws.

Removing all of the lead nose with a file can sometimes salvage otherwise defective bullets; however, if this is done, the shooter should remember that the bullet was originally designed to have a full lead nose and complete removal may alter the stability enough to change the normal accuracy. In addition, the expansion characteristics may also be affected since the soft lead point was intended to aid reliable expansion. When designing the Nosler Protected-Point bullets, the jacket thickness as well as the lead exposed at the tip were major considerations to insure proper expansion. Test target #4, group A shows the average group performance of the Nosler .30-165 grain Solid Base with the entire lead nose removed by using a file. Group 4-A does show that, in this case, there is little effect to the overall pattern size; yet, there does appear to be a tendency to throw a flier, which is probably the result of a lack of consistency between the mutilations.

To view the effects of approximately half of the nose removed from one side of the .30 caliber bullets, .7 grain was removed from the noses of five 165 grain Solid Base bullets. Test group 4-B (target #4, group B) can now be directly compared with test group D on targets #1 and #2. The .30 caliber bullet appears to have reacted in a similar way to that of the .22 caliber bullets. From this comparison it is reasonable to assume that any weight removed from one side of the bullet's nose can be very detrimental to its flight. On the other hand, if the weight is removed equally around the axis of rotation, as with removing the entire nose, the effect on accuracy may be substantially reduced.

We have devoted much of this chapter to point mutilations because they are the most common form of exterior problem; yet, the bases of the bullets may also be important due to the possibility of damage caused by rough handling, etc. We mutilated two groups of bullets to show that, in this case, even major marks or dents seemed to have little effect on the bullet's grouping ability. To test fire group "C" on target #4, five bullets were sanded to destroy the precise intersection between the body of the bullet and the boattail; this was purposely done in an irregular fashion to create the worst possible boattail condition. The group was surprisingly tight and extensive retesting only yielded the same trend. One answer to "acceptable" groups coming from such a situation is the overall quality of the rifle barrel

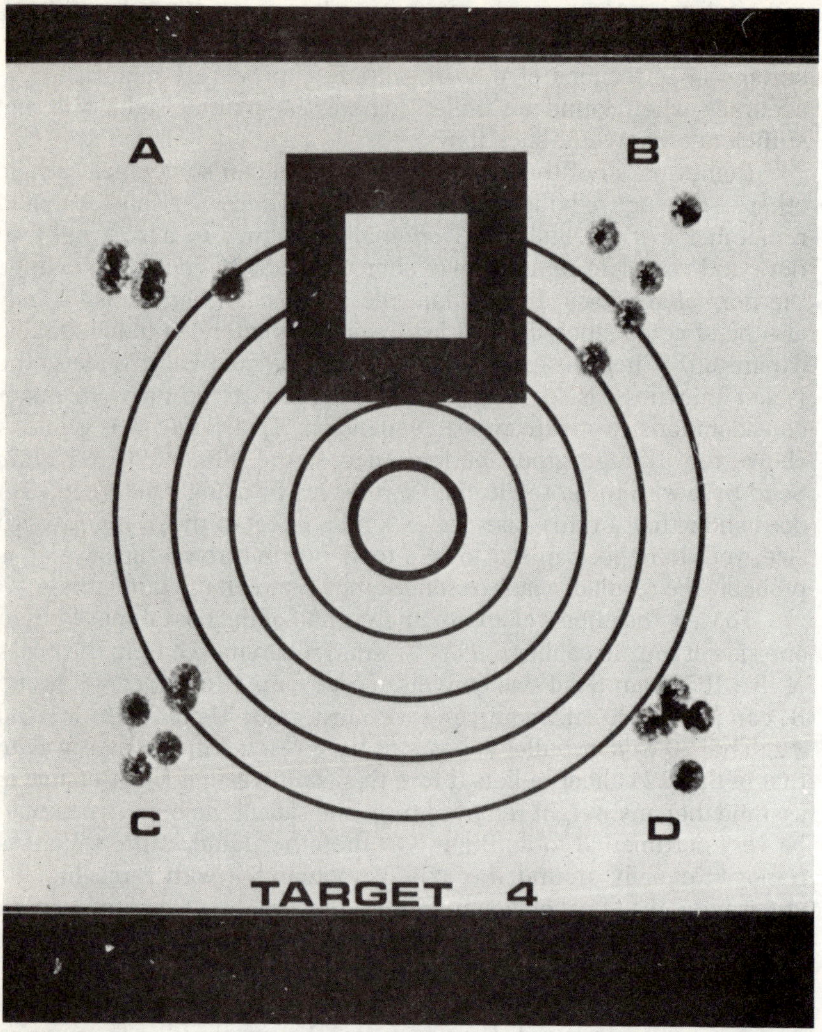

Shown here are .30 caliber groups fired with the entire lead nose removed (group A) .7 grain removed from one side of the lead noses (group B); the entire area of the intersection between the body and boattail altered by sanding (group C); and a large spot sanded on one side of the intersection between the body and boattail (group D).—Photo courtesy of Nosler Bullet Company

coupled with the natural stability of the bullet. The test rifle used to conduct this test series was in new condition with the muzzle riflings well defined. This provides the best possible "launch" for the bullet by

.30 caliber 165 grain Nosler Solid Base bullets used for tests 4-A and 4-B (target #4). Left: *bullet with the entire lead tip removed.* Right: *bullet with about half of the lead removed from one side of the tip.*—Photo courtesy of Nosler Bullet Company

reducing the probability of deflection caused by the muzzle blast. If the test would have involved a bullet capable of match competition, we would have certainly spoiled its ability to shoot in world class competition. However, since the test bullet is a hunting bullet, the effects of such mutilations do not seem to be a major influence on its normal group size.

Group D shown on target #4 was made with bullets that had a flat spot sanded on the "break" of the boattail (the intersection between the body and the boattail). As with group 4-C the pattern was pretty good, again showing the capability of both the rifle and the bullet to overcome this type of mutilation.

There are a number of conclusions which can be drawn from the experiments presented in this chapter; however, without presenting several additional pages devoted to ballistic theory, the following

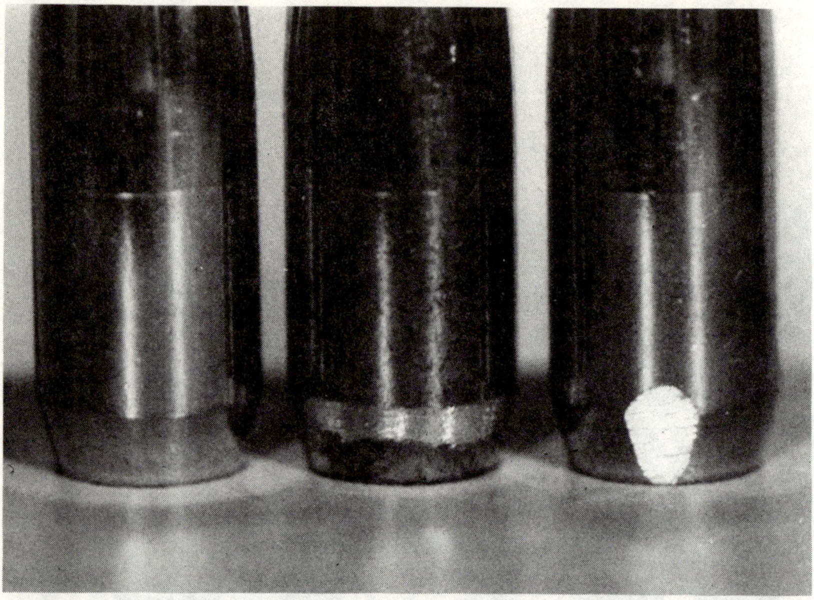

The bullets used for groups C and D of test target #4 are shown for comparison with a "normal" Nosler boattail design (left).—Photo courtesy of Nosler Bullet Company

guidelines may be helpful in predicting the performance of "mutilated" bullets in your rifle:

* Mutilations to the bullet nose *will* cause a change in ballistic coefficient.
* Metal (weight) removed from one side of the axis of rotation will usually destroy accuracy—especially at higher velocities.
* Where accuracy is concerned, mutilations to the bullet base are not as significant as those found on the noses of soft point spitzer designs.
* The effect of any mutilation on accuracy can depend greatly on the rifle.

A benchrest shooter competes on the premise that *all* bullets will have minor imperfections, no matter how small. By combining a slow twist rate with a moderate velocity most bullets improve in overall accuracy. "Imperfections" that are much too small to be detected may cause the accuracy to deteriorate when exposed to a faster twist rate barrel, or higher velocities. As we discussed earlier, the quality of the barrel is also an essential part of the overall accuracy of mutilated bullets. Older, worn, or abused barrels will not tolerate mutilations as well as those with crisp riflings which have been kept in peak condition.

All of the tests performed in this chapter have shown only the accuracy of the particular mutilation. In these tests we did not fire mutilated bullets in the same group with normal bullets because, when working with top quality rifle barrels, some of the mutilations would have printed with the "good" bullets. This would not allow the serious shooter or experimenter to accurately rate the potential performance of the mutilation. By comparing the accuracy performance of the mutilations with that of the "control group" for each bullet, a conclusion may be derived about the potential behavior of a mutilation accidentally mixed in your rifle magazine.

Experimenting with the endless variations of the basic mutilations presented in this chapter will open a new door to the serious shooter or experimenter. The same test series that caused Nosler to introduce the "Protected-Point" design a few years ago may also lead other experimenters to find additional improvements to the basic bullet styles common in modern rifle shooting.

Chapter 32

B&A Dies

While there were many makers of swaging dies from the 1920s to the early 1950s, one particular style of swaging die design was developed and brought to perfection, along with the principle of "swaging up," by a company whose products set the standards for most of today's work. The company was actually an association of two friends, working after their normal jobs and during weekends, to provide precision shooters with the very best possible bullet-swaging equipment for the times. The company was Biehler and Astles, known as B&A Dies.

To give some idea of the company's fame and an insight into the diemaking business in the late 1950s, we have taken directly from a letter Ray Biehler and Walt Astles mailed to their customers in July of 1957 the following quote. The B&A company is no more, and has been out of business for many years, but the development work which they did deserves recognition in any complete book about bullet swaging.

"Letter of Explanation"

July, 1957

Dear Fellow Gun Nuts:

At some time or other during the past several months, those of you receiving this letter wrote in to us for literature and info concerning

B&A bullet swaging press, top view.—Photo courtesy of P. H. Sydney, Penfield, New York

our bullet making dies; in fact, some of you have written more than once. So far, none of you have received any reply. You are all entitled to the most abject apologies from us, if indeed, there are any abject enough to cover the situation! You are also entitled to know, and before getting any further involved with us, should know the reasons

B&A bullet swaging press, side view.—Photo courtesy of P. H. Sydney, Penfield, New York

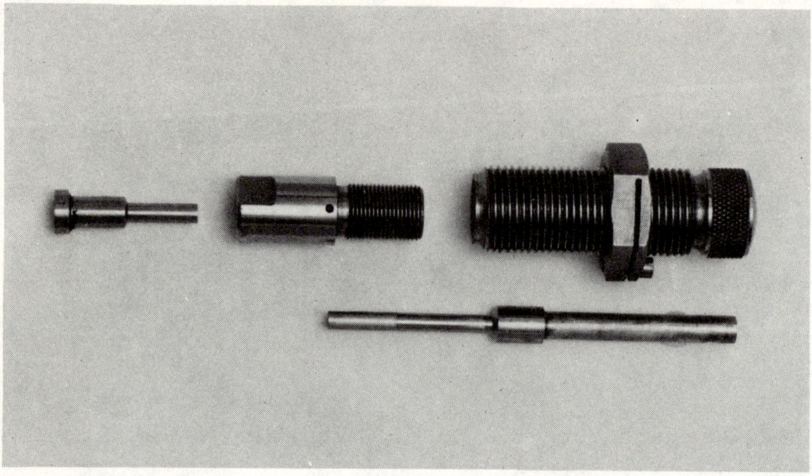

B&A Core Swage Die.—Photo courtesy of P. H. Sydney, Penfield, New York

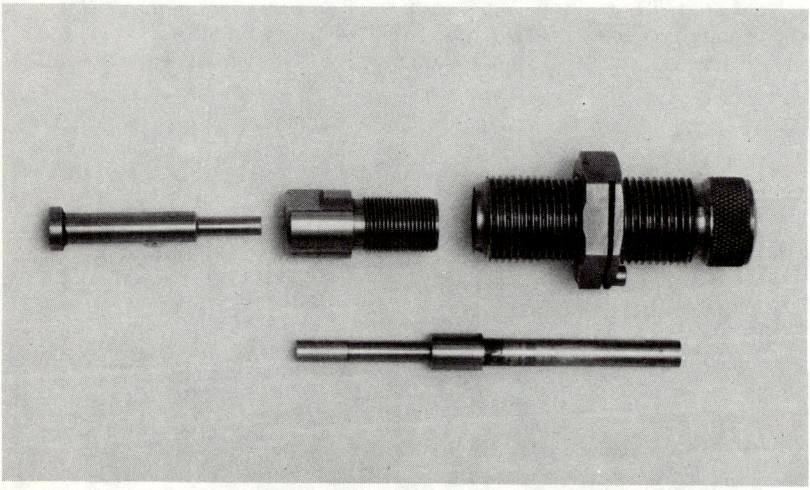

B&A core seat die.—Photo courtesy of P. H. Sydney, Penfield, New York

for the treatment you have received. There are several of them. Here they are:

(1) Nearly a year ago, we ran out of the printed literature we had. Because this literature was rather old and out of date, we wished to do a new write up, and get out a completely new catalog, involving new photographs and additional information. We still have not been able to shake loose long enough from all the

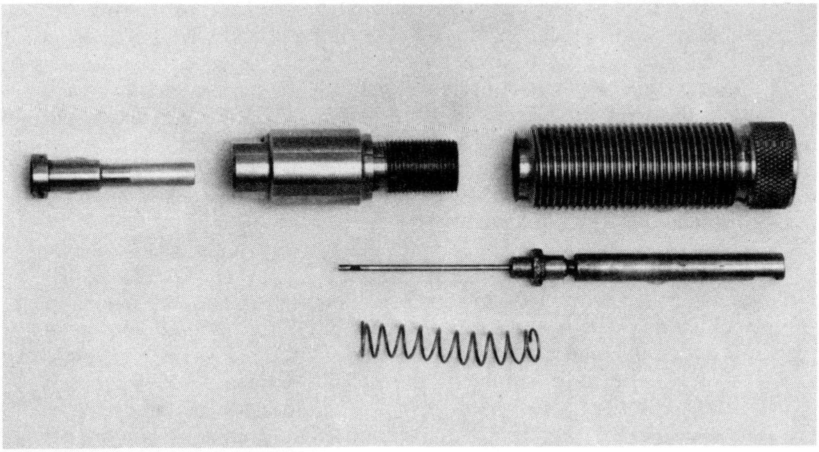

B&A point form die.—Photo courtesy of P. H. Sydney, Penfield, New York

harassing load we're under to complete the new catalog job. We do have a first version of the new write up done, however, and in desperation have had this write up mimeographed, as a temporary measure until the printed catalog is ready.

(2) In spite of our having worked to the very limit of our free time and endurance, we simply have not been able to fully cope with the situation we're in. Much as we love to do this die work of ours, it is, and must remain, a part-time affair. We could never hope to make a living from it on a full-time basis. Even as it is, our dies are and must be expensive. But if we had to charge fully, in accord with the enormous amount of work that goes into them, they would be priced clear out of reach of all but a very few. Thus, our die work began as, and has reamined far more of a hobby to us than a "business" in the usual sense of the word. Everything in connection with both the production and the terrific correspondence problem, must be done evenings and weekends. As things have developed, this is nowhere near enough.

(3) During the past few years, the gods of fortune have seen fit to allow bullets made in our dies to practically "clean house" at the various bench rest matches all over the country. They have accounted for at least 80% or more of all prizes won, and hold all of the world aggregates at 100, 200, and 300 yards.

The end result is that we continually have been simply swamped, clobbered and buried under far more than we could properly take care of. To further complicate things, this past spring, both of us have lost a considerable amount of time—one due to a lot of unavoidable overtime work on his "regular" job, and the other due to an injury to the right hand in the shop, which left same utterly useless in a plaster cast for 5 weeks.

So much for excuses and explanations. Now on to those of you who still may wish to place orders for dies. None of you wishing to do so will be refused. We wish all of you to know, however, that at the present time, we are running a year to 15 months behind on a majority of our deliveries. Anyone placing an order can expect to eventually get his dies, but because of the situation here, we cannot make any definite promises as to just when delivery will be made. All we can promise is that we will continue to do the very best we can for all who are waiting for dies from us.

Insofar as possible, all orders are filled in the order in which they were received. Each fellow must take his proper turn. Occasionally, we receive an order for some odd sized die that we happen to have in stock, and which is not ear marked for any previous order. In these few cases, delivery can be made at once. These cases are, however, very few. There are some items, such as extra die bodies, punch holders, ejection frames and certain others that we usually have in stock, and can make prompt delivery on them. Complete, new sets of dies are the big problem, particularly the bullet swage dies which take such a very great amount of meticulous labor to make.

Now then, to those of you who in spite of the situation here may still want to place orders. We are well aware of the fact that some of you quite likely would have placed orders long ago had you heard from us in reply to your inquiries. The problem is how to be entirely fair to you fellows in regard to a "spot" on the waiting list. We offer this solution. If any of you who have written in still wish to place orders with us, we will date those orders on our books with a date that will be one month later than your first inquiry for information. We will hold this offer open for the next 6 weeks. Be sure to mention that you are one of those who wrote in earlier, and hence have received this letter.

It is to be understood that placing an order will not obligate you in any way. We repeat, you will not be under any obligation. Due to the long waits that are involved, fellows with orders in are not asked to accept delivery when their orders are finally ready, if at that time they no longer want them. In that case, their dies will simply be used to fill some other waiting order with. The main thing, if you do want dies, is to get a spot on the waiting list. All we ask is that you do not ask to be put on the list unless you are as sure as you can be now, that you will want the dies when ready.

Payment for dies. Where probable delivery is so far ahead, under no circumstances do we want any money sent in until we ask for it. Our policy is as follows: About ten days ahead of when we are certain a given order will be ready, we notify the customer that he can then send in his money if he still wants the dies. If so, shipment is then made.

Sincerely,
Ray Biehler
and
Walt Astles

The letter from B&A was followed shortly by the end of the business, although B&A did go on to develop a very fine horizontal stroke, automatic ejecting swage press and actually produced a handful of these before closing their doors forever. Additional historical information about the famous diemakers follows, in their own words, from the text of the information sheet sent under the simple title of "B&A Bullet Making Dies."

> The B and A bullet making dies arose out of a double interest we've long had in the maximum possible accuracy in rifle shooting, and in the bench rest shooting movement, which by its very nature is so much concerned with accuracy. While much interested in all of the factors that can affect accuracy, they are of such variety that very few shooters can hope to indulge in any really extensive investigation of all of them.
>
> In our own case, we had long been particularly interested in the role played by the bullet in the overall accuracy equation. As any experimentally minded shooter soon learns, we too had become aware of the very great differences in accuracy that can exist between different types and quality of bullets. In this connection, along with various other shooters during the early days of the bench rest movement, we had experimented with home made bullets produced in several different makes of bullet making dies available at the time. In this work we had been much impressed by the fact that in spite of what appeared to be various shortcomings in much of the equipment and materials available, it was often possible to produce more accurate bullets than anything we could buy. The good results obtainable, in spite of the seeming shortcomings, suggested very strong possibilities of still much better results that might come out of a determined effort to improve bullet making equipment, materials and methods.
>
> We were most fortunate to be in a position to fully undertake such an effort. We both had considerable toolmaking experience, and had the facilities of a very complete tool shop available to us. There followed a two year period of intensive study, development, and experimentation with different kinds of bullet making dies, bullet making methods and types of bullets. In this work we were very considerably helped, and owe a great deal of thanks to a number of other shooters who were particularly interested in the work we were doing. To such fellows as Jonas Hallgrimsson, Sam Clark, Harvey Donaldson, Col. Townsend Whelen and others, much is owed for their many suggestions, valuable ideas, and the innumerable shooting tests they made of bullets produced with various dies and methods.
>
> It was out of this work, and continuing work since then, that the present line of B & A bullet making equipment arose. Making bullets with this equipment involves four operations, for each of which there is a special tool or die:

(1) Cutting off lead core slugs from commercially available lead wire with a cutter made for the purpose.
(2) "Forming" the cores in the Uniform Weight Core Forming Die.
(3) "Seating" or expanding the cores in the jacket cups.
(4) Forming the points on the bullets in a point forming die, or "bullet swage" as it is usually called.

The materials used in making bullets consist of lead wire and copper alloy jacket cups. Both the extremely high precision and the economy that is possible in hand made bullets has made home bullet making sufficiently popular so that both lead core in various diameters, and bullet jacket cups in practically all calibers are commercially available.

(1) Cutting Cores. The rough sketch . . . illustrates the core wire cutting tool. The lead wire is fed in through a guide tube until it strikes the adjustable stop. The core slug is then cut off by depressing the "knife" handle shown. Bolt holes are drilled in the body of the cutter so that it may be fastened to the work bench.

(2) Forming the Cores in the Uniform Weight Core Forming Die. Although slugs of uniform length can be cut off with the wire cutting tool, the lack of highly uniform diameter in much of the lead wire often leaves something to be desired in the weight uniformity of the slugs.

It is one of the two features of the core forming die to correct this matter. In order to allow this feature to operate, the core slugs are cut off to a length that leaves them just slightly heavier than the finished cores are to be.

The core slugs are inserted in the bottom of the die, and then forced on up in by the base punch, when the punch holder mounted in the loading tool or bullet making press is raised. The pressure of the punch on the core causes the slight excess weight deliberately left in the cutting operation to be extruded off through the three tiny bleed holes in the side of the die. Cores are ejected from the die by pressing on the ejection plunger, and as they are ejected, the extrusions are cleanly sheared off. Since cores made in this way must come out of the die with equal volumes, they also come out with a very high degree of weight uniformity. In cores for a .22 caliber bullet, a maximum, total weight variation of less than 0.1 grain is easily maintained. In larger caliber bullets such as 180 gr. .30 caliber, the variation can be held to approximately 0.15 grain.

The second function of the core forming die is to preform the cores to be a very close match or fit for the inside of the jackets they are to be used in. To accomplish this, the forming cavity in the die is made to correspond exactly in size and shape to the inner contour or shape of the jacket cups the cores are to be used in. The nature of this inner shape or contour varies considerably in different makes and calibers of jackets. In .22 caliber, the Kampen and R.C.B.S. makes have a straight, cylindrical inner shape, whereas the Sierra make has a taper running all the way from the mouth to the bottom, the diameter at the bottom being about ten thousandths of an inch smaller than at the mouth.

In the larger calibers, nearly all of the available makes have a "double" inner shape. This begins at the mouth with a tapered section which extends downward for about one-half to two-thirds the inside depth, and then continues on with a cylindrical section to the bottom.

The preforming feature of our core forming dies is a direct result of an investigation into the perfection of contact between the cores and jackets of bullets that had been made in different ways. By just swaging through the jacket lengthways on opposite sides of a bullet and across the bottom, the two halves could be stripped away, allowing the core to be examined.

The investigation began with bullets in which raw slugs just as they came from the wire cutter were used for core, as was the usual practice at the time. Examination of these cores after stripping away the jackets showed that unless extremely heavy core seating pressures were used in the seating operation, very noticeable contact imperfections were frequently found down around the extreme bottoms of the cores and the lower 1/4 to 1/3 of their lengths.

Perfectly contacting areas had a smooth, glossy appearance, while poor contacting areas were revealed by the presence of dull patches, voids and wrinkles due either to trapped air or other surface defects that did not get ironed out. The sloppy fit of the raw core wire slugs in the jackets made them difficult to fully seat out into perfect contact with the walls of the jackets. Due to any such sloppy fit, the core must undergo considerable upsetting and buckling on being seated with consequent formation of various voids and wrinkles. The resulting action was then a lot like trying to fill a hole in the ground tightly with loose dirt by filling the hole clear full and then tamping on top. The "packing" of the dirt may be very hard, or firm at the top, but will be still relatively loose down deeper in the hole. As mentioned, the same sort of thing had been found in the case of poorly contacting bullet cores. The contact imperfections were either all found, or at least were far greater down around the lower ends of the cores.

The above experiences suggested a pre-formed core shaped with firm pressure in a special die to be a close fit for the insides of the jackets. Being already smooth, and a close fit for the jackets, such cores should squeeze out all but the faintest film of air between themselves and the jackets on insertion. Further, there should be no question of even relatively light seating pressure being sufficient to expand them out everywhere into perfect contact. Tests made with such pre-formed cores showed that they did work out in exactly this way. It was then realized that by the addition of one or more tiny bleed holes in the core forming die, the additional advantage of extreme weight uniformity in the cores could be had at the same time.

(3) Seating the Cores. After the cores have been inserted in the jackets with the fingers, it is the purpose of the seating operation to "seat" or expand them out to an everywhere fully contacting and locked fit in the jackets. The core seating die is used for this purpose. The jacket is placed with its mouth over the end of the base punch, which is then raised up into the die. The die is adjusted in the loading

tool so that when the punch is at the highest point of its travel, it presses on the core with sufficient pressure to firmly seat it.

Special features of B&A core seating dies: First, and foremost, is a development that came out of our experimental work which has come to be known as the "expanding up" principle of bullet making, and which is now generally accepted as an absolute necessity in making bullets of the very highest precision. The purpose of this principle is to eliminate all possibility of any bullets coming out with loose cores due to "spring back" of the jacket material. Following this principle simply means that the jacket cups started with should have a diameter slightly smaller than that of the finished bullets, so that in forming the bullets, the jackets are stretched, or expanded up to the larger diameter.

The point to be remembered here is that while lead is a relatively "dead" material that will stay just about completely "put" when deformed in any way, good jacket materials do not.

Jacket cups containing a proper amount of "temper" for bullet making will always undergo a certain amount of springback towards their former shape and size after being deformed in any way. Hence, in either of the core seating or point forming operations, if the jacket is at all larger than the die it is pushed into, then the danger of more or less loose cores can be expected.

On the other hand, if the jacket is expanded outward in either or both of the seating and point forming operations, then tight cores are a certainty. Under these conditions, springback becomes an advantage rather than a disadvantage. The jacket, being under constant tension, is forced to grip the core with positive firmness.

In our early work with the EU principle, it seemed logical to have the total upward expansion of the jacket about equally divided between the core seating and point swaging operations. In other words, the jackets, the seater and the bullet swage were related so that the jackets were a bit smaller than the seater, and so that the seater was a bit smaller than the bullet swage. Many tests of bullets made in this way conclusively proved their definite superiority over non-EU made bullets. Accordingly, for a while, our dies were made to correspond to the idea of having a certain amount of the EU take place both in the seating and the point swaging operations. However, later and more extended work with the EU principle left us convinced that still better results were to be had when practically all, if indeed not every bit of the EU was made to take place in the seater.

Continued experiments indicated that the ideal situation was to have the jackets come out of the core seater with a diameter lying in the narrow range of from one ten thousandths under the diameter of the swage, on up to, but nothing over, the full diameter of the swage. This discovery of the increased importance of the exact diameter of the core seating die led to the need of the closely "matched" sets of dies which we now make. Our dies are made on a semi-custom basis, and in all of the calibers we make dies for, we make them to yield bullets of any exact diameter requested. In each case, whatever the diameter of the bullet swage is to be, we supply a core seating die to exactly match.

In addition to making our seaters to exactly match the swages they are to be used with, we also make our core seater base punches in any exact diameter required for whatever make and caliber of jackets that are to be used, and for whatever weight of bullet that is to be made. In order to do a clean, neat job of seating cores, it is essential that the base punch properly fit the jackets. If too small, lead will extrude down around its sides when the cores are seated, and if too large, it will bite into the sides of the jackets. For jackets with a cylindrical inner contour, a single, close fitting punch is sufficient for all bullet weights. Where the inner contour is tapered, however, the weight (and hence the length) of the cores must be considered, since different length cores will come to different points in the taper section of the jacket. For customers who wish to make several appreciably different weights of bullets in tapered wall jackets, suitable extra seater punches can be supplied at modest cost.

Other special features of our core seaters are the closely lapped fit of the plungers in the dies, and in the very carefully controlled exact shape of the cups in the ends of these plungers. The exact shape of these cups has a small, but definite effect on the smoothness of operation of the bullet swage, and on the quality of the bullets.

Special note: Sizing of jacket cups in connection with the EU principle. Experience has indicated that in following the EU principle, the total amount of EU should not be less than one thousandth. Ideally, we like to have around two thousandths; however, even three or four does no harm.

In earlier days of home bullet making, most of the jacket cups available had diameters that were very nearly, or even fully equal to the diameter of the bullets to be made. Accordingly, in order to follow the EU principle, it was necessary to "draw" or size such jackets down to a smaller diameter. For this purpose, we made (and can still supply) a special die called a jacket sizing die. Fortunately, however, during the past few years most makes and calibers of jacket cups have been coming through with diameters small enough to allow them to be used just as they come in following the EU principle. Hence, our jacket sizing dies are seldom needed anymore except for special experimental work or other special purposes.

(4) The point forming or bullet swaging operation. After the cores have been seated in the jackets, the bullets have their points formed or swaged on them by means of the bullet swaging die. The core seated jackets are inserted in the bottom of the die and then pushed home with the base punch by raising the punch holder mounted in the loading tool that the die is held in. The finished bullets are then ejected from the die by pressing on the plunger handle, which causes the ejection pin to press on the nose of the bullet. To provide a bit more information for the sake of those not familiar with bullet making methods, it may be well to mention here that all bullet swaging dies fall into two groups. These are the one piece, and the two piece types of swages.

The swages we make are of the one piece type. In a two piece type, the die is made up of two parts, or sections called the ogive section and the body section. The two pieces are held together by a guide bushing, guide pins, or other means while the bullet is being formed. To remove a finished bullet from the die, its two parts are separated. If the die is properly made, the bullet remains in the body section. It is then removed by pushing it on up through the top end of this section. Both of the two types of swages have certain advantages and disadvantages. The one special advantage of the two piece is that both hollow point and exposed lead tip bullets can be made in it. The two disadvantages are:

(1) Even though the original alignment between the two parts may have been perfect, with extended use the alignment can suffer due to wear in the bushing, guide pins, or whatever means is used to hold the ogive and body sections together.

(2) Nearly all of the two piece swages which have been on the market at one time or another have been rather slow and clumsy to operate. Unless a complicated design, quite expensive to manufacture, is resorted to, it is a difficult problem to produce a two piece swage which will be fast operating, yield high precision bullets, and be usable in a majority of the loading and/or bullet presses now on the market.

The particularly special advantages of the one piece type of swage are three:

(1) Completely perfect alignment is a certainty. In a properly made one piece swage, it is impossible to produce bullets having anything less than perfect alignment between their ogive and body sections, and this alignment will continue, of course, for the entire life of the die.

(2) The one piece type is very fast and easy to operate, and this high speed of operation is had at no expense whatever to the quality of the bullets being made.

(3) The principle of the one piece type allows a reasonably simple design to be followed that will permit the die to be used in a majority of the loading and bullet making presses on the market.

The second function of the core forming die is to preform the cores to be a very close match or fit for the inside of the jackets they are to be used in. To accomplish this, the forming cavity in the die is made to correspond exactly in size and shape to the inner contour or shape of the jacket cups the cores are to be used in. The nature of this inner shape or contour varies considerably in different makes and calibers of jackets. In .22 caliber, the Kampen and R.C.B.S. makes have a straight, cylindrical inner shape, whereas the Sierra make has a taper running all the way from the mouth to the bottom, the diameter at the bottom being about ten thousandths of an inch smaller than at the mouth.

The one disadvantage of the one piece type is that it is not well suited to the production of soft point bullets. In most of the one piece swages that have been made, including ours, only the hollow point, or so-called soft swaged type of bullets can be made. However, to the great majority of bench rest and varmint shooters (whose needs have been of particular interest to us) this limitation is of no concern. Surprising as it may be to some of you reading this, the bench rest shooting game has conclusively shown the hollow point type of bullet

> to be the most accuracte that we have today. Time after time, well-made hollow point type bullets have demonstrated their definite superiority over any other type in bench rest shooting where the very highest possible degree of accuracy is what everyone is striving for.
>
> For the varmint shooter, the hollow point type of bullet is also the ideal one. Thus, in this type, we have the very fortunate combination of very high accuracy for the bench rest shooter, and deadly killing power, as well as the high accuracy for the varmint shooter. During our early experimental work, we were able to develop several different two piece swages which quite successfully met the requirements we'd had in mind concerning precision, ease of operation and adaptability to the most popular makes of loading tools. However, meeting all three of these requirements required the designs of these swages to be rather complicated. In comparison, and with all things considered, we like the one piece type we had also developed so much better that we chose it as the one to go ahead with.

The B&A partnership thus resulted in the development of most of the principles of bullet swage die design that are used today, although some of the terminology has been changed. The references to "hollow point" bullets are actually meant to describe what we now call "open-tip" style. These are not the most explosive type of bullets, but one of the least likely to expand (in comparison to the lead-tip or true exposed lead hollow-point type). Methods of making true hollow-point and lead-tip bullets have been developed for production with one-piece dies. The "expanding upward" or EU principle has become the heart of the design of bullet swaging equipment. The use of three dies, the core swage, core seat, and point-forming die (which B&A called the "point swage" or simply "bullet swage" die), is now standard for quality rifle bullets. The bullet maker of today owes much to the early work done by Biehler and Astles.

Epilogue

The young man slid his tires around the twisting country road, tall forests of fir and spruce blurring into an endless green sea. After having heard so much, for so long, about the famous master shooter and diemaker, he had at last located the unlisted telephone number and managed to make an appointment to visit the secluded workshop and home. The pleasant quiet voice coming over the wires had not seemed so mysterious as all the tales the young man had heard, of strange inventions and wonderful achievements, of the work which involved famous names in all phases of the firearms field and led to developments now commonly known to all shooters. Still, he was very excited about the meeting to take place, and pushed his car relentlessly.

Following the slowly explained directions, he drove on and on through the backwoods until at last he came to a small sign in front of a long dirt driveway with the name of the Old Master hand painted on it. He slid past the entrance, backed up, and turned into the drive. In the back seat were several articles he had written on bullet designs, results of tests he had done, a few examples of his gunsmithing work, and some sketches of ideas he hoped would show his ambition and as yet undeveloped skills. He stopped at the one-story ranch house and knocked. No answer. He knocked again. They he caught the faint sounds of machinery running and turned toward a huge warehouselike structure toward the back of the property, almost surrounded by tall evergreens.

The single sliding door, ten feet high and almost as wide, was slightly open. Leaving the car at the house, the young man walked the fifty yards to the door. As he walked he noticed another sound, the crunch of something dry and brittle underfoot. Looking down, he saw the path was made of a dark brownish red material and he bent over to feel it. The entire path was paved with a thick layer of ancient steel turnings and chips from hundreds of machine operations, rusted into a thick carpet of iron oxide. He smiled. What else would one do with twenty years or more of scraps from a dieworks?

The young man stepped inside the doorway into a dimly lit cavern. He could make out the outlines of dozens of huge machines as his eyes slowly became accustomed to the light. In one corner was an overhead driven milling machine like those in the very old books he had read on machine work, slowly chipping its way through a bar of steel, unattended. Leather belts flapped gently around giant steel wheels worn shiny as mirrors by decades of continual operation. Everywhere there were piles of chips, tin cans full of small parts, bars of steel, and machinery packed elbow to elbow, with just a narrow corridor the full length of the building in which to walk.

A busy power hacksaw wheezed and growled as it attached a bar of steel near the doorway, working with a resolute attention to cutting off three inches of metal with no time for distractions. No one was in sight to supervise all this activity. The shop seemed to be running itself, with a purpose born of long, long familiarity. It seemed as if the Old Master had waved his arm and said, "I'm going to take a little rest; go ahead and finish up for me. You know by now just what needs to be done."

"Hello! Anybody here?" The young man shouted and heard a slight echo from the back of the dark cavern. A slight clacking noise seemed to drift down a very narrow staircase to his right. He took one last long look down the row of lathes and mills and strange machines whose purpose could only be a guess, saw no humans responding, and turned to the stairs. Ech step brought up a little cloud of dust, and every few steps had to be taken over the top of a large can full of odd-looking steel parts, boxes filled with rods or dies or bullet jackets, or piles of ancient tool parts. At the top, a single dim bare bulb could be seen in a far corner, lighting piles of boxes and crowded little rooms built, apparently, as the need for them arose, without a great deal of regard for plan.

There, under the light, was a solitary tall figure, bent over an offset spring press that was cranking out page after page of some small booklet. Piles of crumpled paper were all over the floor; ink was on everything. Metal printing plates were stacked on home-built wooden benches and hung on nails driven into the uncovered framework of the building. The young man approached and noticed the kindly face, the gaunt build and white hair, and then—those penetrating steel-blue eyes. It was a face that betokened a great deal of experience, and intelligence.

"Hello," said the young man, slightly surprised and self-conscious in the focus of the tall man's eyes. "I . . . I called you, earlier. Are you . . . the Old Master?"

The Old Master looked straight at the young man for a moment, then reached over and turned off the power to the printing press. He wiped his hands on his coveralls.

"Why, yes. I suppose I am." Then he handed the young man one of the wet pages in the collection basket at the end of the press. "What do you think of this?"

There, in a little line drawing, with an explanation set in crude typewriter stencil, was the detailed illustration of a remarkably simple method of handling one of the most time-consuming jobs a handloader has to do, with an entirely practical and obviously low-cost tool. The young man read carefully, studied the sketch, and looked up at the Old Master.

"Why, this is great! If you patent this idea and market it, you'll drive half the tools used for this job off the market! This idea is so simple. Did you invent it?"

"No, I don't really invent things," the Old Master said. "I'd rather say it this way: I try to see a need and then just make a tool that fills the need. Invention sounds too pompous." He motioned toward the paper in the young man's hands. "You can keep that, if you like. I'm not going to patent that. I like to make things. I don't like production, anyway. It's boring. I make enough tools for my friends and shooters who hear about me, and they need instruction to go with the tools. That's all this is for, instruction."

They walked downstairs, talking about inventions and selling tools and manufacturing ideas and a dozen other things. The Old Master showed the young man his guns, his machinery, and some of his ideas both old and new. Uncovering one dusty box after another, the Old Master dragged out rusty prototypes of tools which were either in standard production and known world-wide under some other name, or which the world had never seen—and probably never would—because the Old Master had never brought them to the attention of anyone who knew what they were worth. The young man marveled at the tremendous work and clever thought that had gone into these products, and the Old Master grew conscious of a worth-while ambition and appreciation for his kind of thinking.

But the greatest appreciation and interest the young man exhibited was for the Old Master's bullet swage dies. Here was the real claim to lasting notability, in these small tool steel cylinders and polished rods, in the dozens of different systems and methods and tooling and techniques he had so carefully worked out, tested, discarded, improved upon, and offered in small numbers to shooters over the years. Here, thought the young man, was real challenge and adventure—to learn the secrets of diemaking, and be able to produce bullets which could stand as monuments to human skill and intelligence. So many of the young man's own ideas could only be brought to full power through the use of very special bullets no factory had ever dreamed of, and no mold could ever cast.

The day went swiftly, and the Old Master and the young man parted as friends. The following weeks saw the two of them together more and more often, working side by side, the young man offering suggestions and learning gently but positively where he had little to offer, and where he had much to learn. The weeks turned into months, and the months to years. In all this time, the young man had grown to respect the Old Master's strengths and to overlook his weaknesses, and had learned to follow his instructions, to question them and even, at times, find better ways which the Old Master acknowledged with a smile.

Eventually, the young man was making his own ideas into finished steel, bringing life to the line drawn on paper and form to the numbers dialed up on machines. The Old Master no longer worked in his shop; slowly, he had sold his machinery and done less and less with the ideas in his mind. But his methods, his years of experiments and failures, and the techniques born of his successful works, guided the hand of the man who now guided the diamond-loaded laps into the cavities of what were becoming known as the finest bullet swaging dies in the world. The Old Master had seemed concerned at their last meeting, in fact, that the young man didn't quite realize what was happening.

"You have a responsibility. I'm turning this over to you, because there is no one else. You will have to give up much; there is no wealth in it. I would not recommend that you do it. I would point to myself and say, quit now, because the world does not reward my kind of thinking. But the reward is there. It comes with the responsibility." And the Old Master coughed, and looked very pale, but managed a smile.

Many years later, except for a single manuscript treasured by those shooters who understood, there was little enough in the way of a physical monument to mark the long difficult struggle and the many brilliant but esoteric successes of the Old Master. The young man was no longer young. He still toiled over the ancient techniques of diamond dust, brass laps, and tool steels, even though a vast number of new methods had been tested and various parts put to use in one way or another. Electronics, chemical methods of machining, new exotic steels which the Old Master would have enjoyed like a kitten with a new ball of yarn—all these and more contributed to the original skills learned so long ago in the dimly lit workshop in the forest. But in the final test, only the skilled hand touching the tool, communicating with the die, feeling out its form and finish—questioning it, then demanding that it obey a precision beyond unthinking automation, untouched by the robots of mass production—could ever produce the desired quality.

The diemaker stood by his drawing board, pencil in hand. He was alone, and tired. On the paper was a daring new concept in a hunting bullet, which would change the very usefulness of many calibers and extend them into new fields formerly impossible for them to handle. But the paper had no bullet drawn on it. Instead, there were lines representing a set of dies and punches. Only a person familiar with both machine work and bullet swage dies could see the bullet in his mind that these dies would make. The idea had come to the diemaker early this morning, just before he got out of bed.

A knock sounded against the shop door.

"Come in, it's not locked," said the diemaker.

A serious-looking young man opened the door and walked over to the diemaker. The young man was dressed in a suit and tie, and the diemaker was in coveralls. The young man said hello, and asked if the Old Master was in.

"I've got a copy of his book, here, and I wanted to show him some ideas I had, to see what he thought," the young man said, producing a dogeared little paper-covered booklet with the familiar words in it. But before the diemaker could answer, the young man whistled a loud, low whistle and peered down at the drawing.

"Good grief!" he muttered. "I really blew it! I didn't have half the idea, yet, and here's one a generation ahead of it. Whose idea is this, anyway?"

"It came to me this morning," said the diemaker.

The young man looked up in amazement. "Then—then *you* must be—No, it can't—" He looked down at his old copy of the book and up again at the figure standing before him, with the gray hair and white stubble of a beard, worn coveralls and hands wrinkled and permanently marked with the tell-tale grit of years of machine work and soluble oils. "Are you—the Old Master?"

The diemaker blinked. He looked slowly around the room, through the open door that led to the machine shop, where some mills were running in their automatic feed mode, doing slow, careful work to their preset shut-down stops. The sounds of the chip, chip, chipping brought him swiftly backward through all the years, to that first day when his life changed to a single dedication. Some things never change, like the sound of a slowly revolving mill cutter biting its way through soft steel. He thought of how easy and how brilliant the drawing must appear to the young man, and yet he knew that the idea which came to him so clearly was just the final assembly of a thousand failures, the last changes in a million quick thoughts over almost an entire lifespan.

And then, in that same brief moment, it occurred to him also: no, not just a single lifespan but all the work and frustration and triumphs of another lifetime, the lifetime of his teacher. And not just that, but again all the knowledge and effort, and yes, the courage of his teacher's teachers—and all of their teachers, however little or however much their work had been directly aimed at the product that now awaited life on the drawing board. In that moment, with the impact of something so great and timeless and powerful it stirred him to the very soul, he recognized an overwhelming responsibility. Being the best was no longer sufficient by itself.

"Are you the Old Master" The words still echoed softly against the far wall of the shop.

The gray-haired diemaker looked the young man straight in the eyes, and smiled with gentle relief at the intelligence he recognized in them. He could learn easily.

"Why, yes . . . I suppose I am."

Appendix

Melting Points of Alloys of Copper and Zinc

% of Zinc	
0	1084°C
10	1040
20	995
30	930
40	900
50	880
60	820
70	780
80	700
90	580
100	419

Melting Points of Alloys of Lead and Tin

% of Tin	
0	326°C
10	295
20	276
30	262
40	240
50	220
60	190
70	185
80	200
90	216
100	323

Approximate Bullet Weights for Jacket Lengths

This table will give the bullet swager some idea of the weight of bullet a given jacket length will make. Because jackets vary in thickness, and ogive length and shape will determine a good percentage of the bullet weight even with the same overall length and caliber, the table is only a rough guide. In calibers over .308, a one-inch jacket length usually makes the standard middle range of weights in a flat-base round-nose bullet.

Caliber	*Length*	*Hollow-Point Weight*	*Lead-Tip Weight*
.224	.600	40 gr.	45 gr.
.224	.690	50 gr.	60 gr
.224	.705	53 gr.	63 gr.
.243	.750	65 gr.	75 gr.
.243	.825	75 gr.	85 gr.
.243	.935	85 gr.	100 gr.
.257	.860	85 gr.	100 gr.
.257	.975	105 gr.	115 gr.
.264	.960	115 gr.	125 gr.
.264	1.095	125 gr.	140 gr.
.270	.975	120 gr.	135 gr.
.270	1.095	140 gr.	155 gr.
.284	.995	125 gr.	145 gr.
.284	1.095	140 gr.	160 gr.
.308	.770	110 gr.	125 gr.
.308	.925	130 gr.	150 gr.
.308	1.085	160 gr.	180 gr.
.308	1.125	168 gr.	200 gr.

Ballistics Relationships

w = weight of bullet in pounds
v = velocity of projectile in feet per second
d = diameter of bullet in inches
g = acceleration of gravity = 32.16 ft/sec^2
i = coefficient of form from Ingalls chart
C = ballistic coefficient
$S.D.$ = sectional density
E = kinetic energy of projectile in foot-pounds

$$C = \frac{w}{id^2} \qquad E = \frac{wv^2}{2g}$$

$$S.D. = \frac{w}{d^2} \qquad C = \frac{S.D.}{i}$$

Lead Wire Size for Jacket and Caliber

.224 made from .22 rimfire case	3/16 or swage to .201″
.224 commercial jackets	3/16 or swage to .189″
.224 made from drawn 6mm	5/32
.243 made from .22 rimfire case	3/16
.243 commercial jackets	3/16 or swage to .201″
.243 made from drawn .257	3/16
.257 commercial jackets	3/16
.257 made from drawn 6.5mm	3/16
.264 commercial jackets	3/16 or 7/32
.264 made from drawn .270 jackets	3/16
.270 commercial jackets	3/16 or 7/32
.270 made from drawn 7mm jackets	3/16
.284 commercial jackets	7/32 or 1/4
.284 made from drawn .308 jackets	7/32
.308 commercial jackets, under .95″	1/4
.308 commercial jackets, over .95″	1/4 or 7/32
.308 drawn from 8mm jackets	7/32
.312 or .318 (8mm) commercial jackets	1/4
.338 commercial or drawn 9mm	1/4
.358 commercial or two-part using HG jackets	5/16
.357, .38, and 9mm Handgun	5/16
.375 commercial or two-part using HG jackets	5/16
.41 Handgun, all .400 rifle calibers	11/32 to 3/8
.44 Handgun, .444 Marlin, other .429″ bullets	3/8
.45 Handgun, .458 rifle, 45-70, other similar	3/8

Relationships of Truncated Conical Bullet Dimensions

The following information is very useful for making reamers to cut nose punches for truncated conical bullets, like the Keith style. It does involve the use of some trigonometric conversions, but these can be done on modern scientific calculators or looked up in any good math table book. The angles at which to grind the reamers can easily be determined by using the established amount of shoulder and a nose width at the tip which has been proven effective for the calibers given. By using the formulae given, one can select any other amount of shoulder and nose width and determine the proper reamer angles. The same formulae can be applied to making hollow-point punches on a lathe, simply using the bottom of the cavity as the nose width at the tip, and the maximum cavity size at the tip as the width of the bullet nose at the shoulder.

A = diameter of bullet shank
B = diameter of bullet nose where it joins the shank (shank less amount of shoulder)
C = depth of shoulder between shank and nose (one side in a cross-section drawing)
D = width of nose at the tip (minimum diameter of the nose)
E = length of the nose, starting with the shoulder and ending with the tip
a = angle of slope from start of nose to tip of nose compared to bullet centerline

Chart for Revolver Bullets

A	B	C	D	E	a
.357	.315	.021	.225	.225	10°
.410	.366	.022	.265	.280	10.2°
.429	.380	.0245	.285	.295	9.15°

Chart for Autoloader Bullets

A	B	C	D	E	a
.251	.220	.0155	.1255	.133	19.5°
.309	.270	.0195	.1545	.164	19.5°
.355	.311	.0220	.1175	.188	19.5°
.356	.312	.0220	.1780	.189	19.5°
.452	.396	.0280	.2260	.240	19.5°

$$a = \text{arc tan } \frac{B - D}{2E}$$

Volume Formulae

The formulae for the volumes of various solids bounded by curved surfaces are very handy in determining the weight that a given shape of bullet will have, using only simple algebra rather than calculus. The various shapes of bullets can be broken down into truncated cones, spheres, cylinders, cones, spherical segments, and other shapes which have simple mathematical formulae for volume. The volume can then be multiplied by the density of lead or the various alloys of lead to discover an approximate weight for the bullet.

Volume of a sphere = $4.189r^3$ or $\frac{1}{6}\pi d^3$

Volume of a spherical segment, of height h, radius of sphere $r = \frac{1}{3}\pi h^2(3r - h)$

Volume of a cylinder, r = radius of the base and h = height, $= \pi r^2 h$

Volume of a cone, height of h, and radius of base r, $= \frac{\pi}{3} r^2 h$, or $1.047 r^2 h$

Volume of a truncated cone with radius of base r_1, radius of top r_2, and height of h, $= \pi \frac{h}{3}(r_1^2 + r_1 r_2 + r_2^2)$

Volume of oblate spheroid (solid formed by rotation of an ellipse about its minor axis) where a is major and b is minor semiaxis,

$$= \frac{4}{3}\pi a^2 b$$

Volume of prolate spheroid (solid formed by rotation of an ellipse about its major axes) where a is major and b is minor semiaxis,

$$= \frac{4}{3}\pi a b^2$$

Nominal Calibers and Bullet Diameters, Rifle

There is often quite a variation between different barrels, different makes, and even different cartridges with the same general caliber designation but different specific names. One cannot say that a .40 caliber rifle is always .400, .402, or even .412 inch across the grooves; various calibers of rifles will take different diameters of bullets even though they may have the same nominal caliber in their caliber name. However, more modern U.S. calibers generally have standardized on certain bullet diameters with certain nominal calibers of rifles and handguns. This chart is a guide to the general practices of barrel and gunmakers, while the precise diameter of any barrel should be found by actual measurement of the groove and land dimensions. Do not use this chart as an absolute guide to the size of bullet for a given caliber, but only as a general guide to the relative sizes and their place of order of bullet diameters.

Caliber (Nominal)	*Bullet Diameter (Inches)*
.14	.145
.15	.150
.17	.172
.20	.204
.22	.224
.228	.228
.23	.230
6mm	.243
.24	.243
.25	.257
.26	.264
6.5mm	.264
.27	.277
.28	.284
7mm	.284
.30	.308
.32-20	.312
.303 British	.312
.32 Win	.320
8mm (obsolete)	.318
8mm (current)	.323
.338 Win	.338
.34	.338
.348 Win	.348
.351 Win	.351
.35	.358
9.3mm	.366
.375	.375
.38	.375
.38-40	.401
.401	.406
.44-40	.427
.44	.429
.45	.458

Nominal Calibers and Bullet Diameters, Handgun

Quite often the caliber designation for a handgun will be a very much different bullet size than the same designation for a rifle, as with the .25 caliber. A .25 caliber rifle uses .257 diameter bullets, while a

.25 caliber pistol uses either .250 or .251 diameter bullets. The practice of fitting obsolete handgun calibers with heel-type bullets, in which the bullet base fitted inside the case while the rest of the bullet shank was the same diameter as the outside of the case at the mouth, made some calibers considerably smaller when the later practice of making the bullets fit entirely within the case came into common practice. Thus we now have the .38 caliber being a .357 diameter (usually) and the .41 Long Colt using a .386 bullet (reduced to fit inside the case, instead of gripped by the case mouth on a smaller heel at the base). Some of the more common calibers with their bullet diameters are specified here. When in doubt, always measure the barrel of the particular gun.

Caliber (Nominal)	*Bullet Diameter (Inches)*
2.7mm	.107
3mm	.120
4.25mm	.167
5mm	.202-3
5.5mm	.225
.22 JET	.223
.22	.224
.25 ACP	.250-1
.256 Win	.257
6.5mm	.264
7mm Nambu	.280
7.65mm	.308-9
.32 ACP	.308-9
.32 S&W	.312
.32 Colt	.313
.35 S&W	.309
7.5mm	.317-.325
8mm	.320-3
9mm	.355
.357 Mag	.357-8
.380 ACP	.355-6
.38 Special	.357
.38 S&W	.359
.380 Revolver	.375
.41 Long Colt	.386
.41 Mag	.410
10.4mm	.422

.44 S&W American	.434
.44 S&W Russian	.429
.11mm	.425-6
.44 Special, Mag	.429
.45 ACP, Auto-Rim	.452
.45 Colt	.454

Glossary

Base—the bottom part of the bullet; the portion which comes out of the gun last.

Boattail—an angled portion of the bullet shank which joins the shank to the base, as opposed to a flat-base bullet in which the shank joins the base with a slight radius or an abrupt edge; used on long-range bullets to reduce drag at subsonic speeds, or to increase length while maintaining reasonable weight.

Bottom Punch—the punch which is removed from the swage die to insert cores and jackets; can form either nose or base depending on design, except in the ogive die, where it always forms the base of the bullet.

Bullet—that part of a cartridge which is propelled to the target by the pressure of burning powder gases; an assembly of lead and jacket material shaped by the action of the swaging die.

Bullet Jacket—a deep-drawn cup of heat- and abrasion-resistant metal which is formed over the lead core of a bullet to protect the lead from friction, heat, and to control the bullet expansion. The jacket is the "skin" of the bullet.

Bullet Swaging—a method of producing bullets by forming the bullet materials under high pressure within a die, or in several steps using various dies to control diameter and shape.

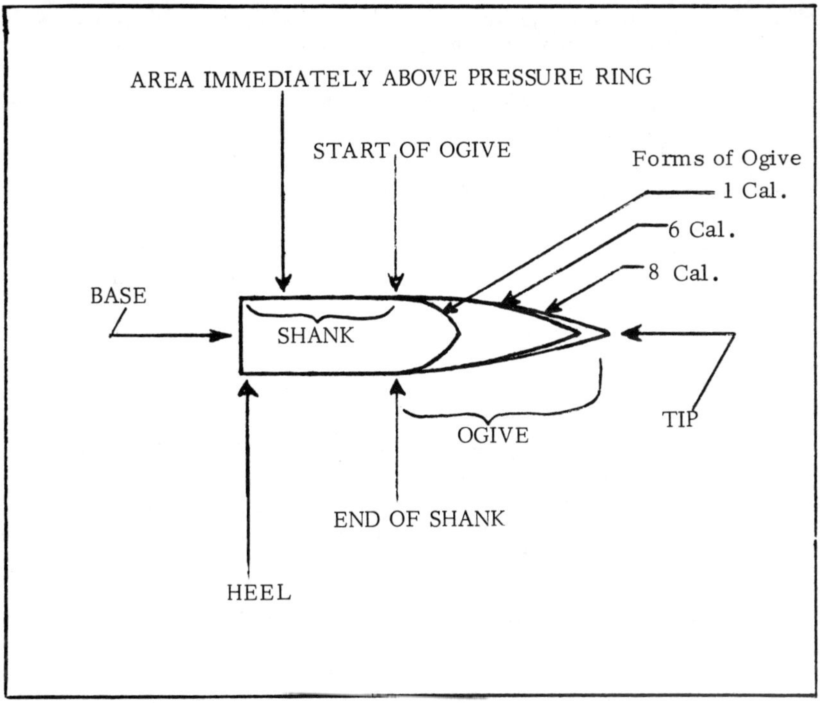

External parts of a bullet.

Caliber—the nominal groove-to-groove diameter of the barrel in which a bullet will be fired; the common name for a cartridge or bore size (which may have nothing to do with actual bullet diameter).

Calibers of Radius—measurement of the curvature of a spitzer bullet ogive; length of radius of an arc, using the caliber of the bullet as the unit of measurement, drawn from a point perpendicular to the end of the shank and beginning of the ogive, the curve of which is the ogive of the bullet.

Cannelure—indentation rolled around the circumference of a bullet to serve as a crimping groove, to aid in identification, or to control jacket expansion or prevent core separation.

Core—the filling of a jacketed bullet, made of soft lead; may be cast in a mold or cut from rolls of lead wire.

Core-seating Die—straight-walled die which is fitted with one bore-sized punch and a removable punch of diameter determined by bullet style and jacket size, for the purpose of seating the lead core into the jacket.

Core-swaging Die—straight-walled die with small bleed holes in its side, through which excess lead from cut or cast cores can be extruded to bring cores to uniform size and weight; also known as a "bleed die" or "squirt die."

Cup—another term for jacket, generally used only for jackets which are approximately the same length as their diameter, or shorter. Also, a partially drawn jacket that is not yet completed.

Cup Base—depression in the base of a bullet to allow the base to expand upon firing, thus filling the grooves of the barrel even if the bullet is undersized. Also used to reduce weight while keeping bullet the same length.

Cup Nose—a large cavity-nosed bullet used in handguns to give rapid expansion at low velocity.

Ejection Punch—the small-diameter steel pin punch that pushes the bullet out of a point-forming die by its nose; determines the minimum point closure of the bullet.

Ejection Ram—a steel rod placed in some kinds of bullet swage dies to push down the ejection punch or the top punch, and thus eject the bullet material. Not used in automatic ejecting presses.

External Punch—any punch which can be removed from the die to insert bullet material; comes completely out of the die during ejection; the "bottom punch" in some die sets.

Full Jacket—a bullet style in which the base is open and the tip fully covered by the jacket, as with military bullets; a jacket long enough to cover the entire core of the bullet. Designation will change depending on the style and length of core used with a given jacket.

Gilding Metal—an alloy made with 5 to 10 percent zinc and the balance copper, used to make modern bullet jackets.

Half Jacket—a jacket which covers only about half the shank of a bullet; a handgun bullet so made. Used only in moderate to low-velocity handgun loadings.

Internal Punch—any punch made to fit the top of a die, which remains in the die during all operations and is used to form one end of the bullet and to eject the bullet material; also called the "top punch" in some die sets.

Ogive—the curved nose section of a bullet, beginning at the shank's ending point and ending at the bullet tip.

Point—the nose and tip of a bullet; all of the bullet forward of the shank.

Point-forming Die—a semiblind cavity die with cavity shaped exactly like the bullet it will form, used to bring the nose of a seated core to proper ogive form. Required any time a bullet must have a smooth stepless transition from shank to nose, and used with a core-seating die in all cases.

Power Ejector—mechanical device which multiplies the force which the operator applies by hand to a lever, for ejecting bullets from certain kinds of bullet swage dies. Not used with automatic ejecting presses.

Rebated Boattail—a boattail joined to the shank by an abrupt step or small radius, rather than directly joining the shank.

Rimfire Jackets—bullet jackets made by drawing a fired rimfire case to proper shape in a draw die.

Shank—the full-diameter, parallel-sided portion of the bullet; the section of bullet between the base and the ogive.

Three-quarter Jacket—a bullet jacket which covers substantially all of the shank; designation changes with various weights and hence core lengths used with a particular jacket.

Tip—the extreme end of the point of a bullet, opposite from the base.

Top Punch—the punch which remains partially within the die during all operations, and is pushed down to eject bullet material.

NOTES

NOTES

NOTES

NOTES

Born in Oklahoma, DAVID R. CORBIN currently resides in Medford, Oregon, where he founded what is presently the world's largest manufacturer of bullet swaging equipment. Trained as a communications and electronics engineer, his main interest has always been experimental firearms investigation, and with the development of his company, his one-time hobby has become his career. In addition to his business pursuits, he is a columnist and feature writer for several special trade journals; he is the author of *The Corbin Handbook*, and he co-authored *The Bullet Swage Manual*. Corbin is a life member of the National Rifle Association.